Water and Sanitation in the New Millennium

K.J. Nath • Vinod Prakash Sharma
Editors

Water and Sanitation in the New Millennium

Springer

THE NATIONAL ACADEMY OF SCIENCES
राष्ट्रीय विज्ञान अकादमी
ESTD 1930
भारत
INDIA

Editors
K.J. Nath
Former Director, All India Institute of Hygiene and Public Health
GOI, Kolkata, India

Vinod Prakash Sharma (deceased)
Centre for Rural Development and Technology
Indian Institute of Technology
Hauz Khas, New Delhi, India

ISBN 978-81-322-3743-3 ISBN 978-81-322-3745-7 (eBook)
DOI 10.1007/978-81-322-3745-7

Library of Congress Control Number: 2017950011

Printed on acid-free paper

This Springer imprint is published by Springer Nature
The registered company is Springer (India) Pvt. Ltd.
The registered company address is: 7th Floor, Vijaya Building, 17 Barakhamba Road, New Delhi 110 001, India

Foreword

The global importance of water, sanitation and hygiene for development, poverty reduction and health is reflected in the United Nations Millennium Declaration, in particular its eight Millennium Development Goals, in the reports of the United Nations Commission on Sustainable Development and at many international fora.

In India, the successive governments launched several flagship programmes like the National Mission for Clean Ganga, Jawaharlal Nehru National Urban Renewal Mission, National Rural Drinking Water Quality Monitoring and Surveillance Programme, and, more recently, the Swachh Bharat Mission and Smart City Mission. However, the planning and implementation of these programmes and methodologies/technologies adapted need critical review and evaluation. The National Academy of Sciences, India (NASI), being a premier organization of scientists, engineers and health professionals, is seriously concerned about the prevailing situation in respect of community water supply and sanitation practices and its very critical impact on community health. We sincerely feel that it is necessary for the leading scientists, technocrats and health professionals and the sector leaders to sit together and deliberate in depth about the maladies and seek remedial and corrective steps. In the above context and as conceptualized and planned by Dr. V.P. Sharma and Prof. K.J. Nath, NASI organized the **1st Policy Support Brainstorming on Safe Water and Sanitation for Rural and Urban India** in collaboration with WHO and Sulabh International Social Service Organisation (SISSO) in July 2009. It was a highly successful event attended by distinguished scientists and technical experts from all over the country. Along with the **Allahabad Declaration**, a set of precise and scientific recommendations were adopted which were forwarded to all the concerned departments. As a follow-up of the brainstorming, participants formulated a number of research and development proposals which were forwarded to the Department of Science and Technology. Further, for an in-depth deliberation and review of the progress in the sanitation, water supply and environmental management sector in the context of the policy and programme of the government, the academy organized **the 2nd National Brainstorming in Allahabad**, on 20–22 September 2012, in collaboration with SISSO. The recommendations of this event were particularly contextual in the sense that a comprehensive policy initiative was suggested at the national, state and local levels.

Dr. V.P. Sharma (who left us in October 2015 for the heavenly abode) was very much concerned about the continuing crisis in the water and sanitation sector in

India. Prof. K.J. Nath and Dr. Sharma deliberated on the issue and proposed the 3rd Brainstorming in November 2014 which was approved by the council and organized successfully. Meanwhile, it was felt that the papers presented during the 2nd Brainstorming should be published for larger dissemination among the scientists and professionals in the country. Altogether, 24 papers were presented in 7 thematic sessions. Issues and critical challenges facing the country, on various aspects, were discussed during these sessions. This book is the compilation of those presentations in a systematic manner, so that the valuable information/data generated out of the deliberations will be stored for helping the policymakers and programme implementers in a scientific way.

I am very much grateful to Prof. K.J. Nath, the late Dr. V.P. Sharma and all other contributors for this valuable treasure. The contributions of the distinguished authors are of critical importance and would be extremely useful for the scientists and professionals, working in the sector, to engage in further deliberations and suitable actions.

NASI
Allahabad, India

Akhilesh K. Tyagi

Preface

World Health Assembly 1998, Alma-Ata, adapted four key strategies for attaining health for all. One of these key strategies was promoting healthy lifestyles and reducing risk factors to human health that arise from environmental, economic, social and behavioural causes. If the agenda has remained unfinished by a wide margin, the primary reason could be found in our failure to develop an enabling policy for promoting a hygienic environment conducive to healthful living. Environmental services such as community water supply, sanitation, control of air and water pollution, waste disposal and personal and domestic hygiene along with nutrition and health education are central to the concept of preventive and social medicine, and they are the key pillars of primary health care.

Access to safe water is now a basic human right, as per the resolution of the 29th Session of the UN Committee on Economic, Social and Cultural Rights, November 2002. To protect human health and to prevent sickness and mortality, community water supply needs to be reliable, in sufficient quantity, of adequate quality and readily accessible to all segments of the consumers. The people in the developed countries are mostly having the privilege and opportunity of having adequate quantity of water of acceptable quality, anytime and anywhere in the country. The same is not true for the citizens of most third-world countries like India, particularly the poor. Water scarcity, surface and ground water contamination and lack of access to safe drinking water by the poor are among the main obstacles to full enjoyment of the right to water in our country. The health outcome of the lack of sanitation and safe water is enormous, globally 4 billion cases of diarrhoea, 2.2 million deaths per annum and 62.5 million disability-adjusted life years (DALYS) lost. The World Bank estimates that 99.9% of deaths attributed to poor water supply, sanitation and hygiene occur in the developing countries. An estimated 60–80% of all diseases and over 1/3 of deaths in the developing countries are caused by environment-related factors, and on an average as much as 1/10 of each person's productive time is sacrificed to environment-related diseases.

Following the Alma-Ata and Mardel del Plata conference, the International Drinking Water Supply and Sanitation Decade was launched in India with high hopes and expectations for supplying safe and potable water to most of our population in the 1980s. Unfortunately, the decade has been for all practical purposes a story of misplaced priority, lost opportunity and unfinished agenda. It is time for a serious introspection to identify the maladies and plan for the future. The scientific and technical leaders of the country as well as the policymakers in the national government owe it to the community to explain our failures during all these years since independence. It is also extremely urgent that a blueprint and a road map for providing safe, clean and affordable water to all segments of our community living in urban, peri-urban and rural areas need to be drawn up, particularly in the context of the new programmes and initiatives being launched by the national government.

It is in the above context that NASI organized the 1st Policy Support Brainstorming on "Safe Water for Rural and Urban India" in Allahabad, July 2009, in collaboration with WHO and SISSO. It was a highly successful event, attended by the leading scientists, engineers and sector leaders of the country. At the end of the two-day brainstorming, the Allahabad Declaration was adopted, along with a set of concrete recommendations. As a follow-up, a number of R&D proposals were also formulated and submitted to the Department of Science and Technology, GOI.

I would be failing in my duties if I don't mention the total support, advice and guidance that I received from the late Dr. V.P. Sharma, the joint editor of this book, all through the three brainstorming programmes on safe water and sanitation. Dr. Sharma, who was an internationally acclaimed entomologist, vector control expert and malariologist, had been a great source of inspiration and guidance for me during the last four decades. We worked together in the WHO global team for the in-depth evaluation of the Indian malaria control programme in 1985, and ever since then he had been a friend, philosopher and guide for me. It was his concern for environmental issues including sanitation, waste management and provision of safe water that stimulated the idea of the 1st Brainstorming. But for the committed support and leadership of Dr. Sharma, the successful organization of the three policy support brainstorming programmes would not have been possible. His untimely death has been a great loss for the scientific community of the country and a personal shock for me. I am sure that the president, NASI and council members along with the fellows and members would join me in dedicating this book to the memory of Dr. V.P. Sharma.

Dr. Sharma and I deliberated at length regarding the continuing crisis in the drinking water and environmental sanitation sector in the country and decided to organize the 2nd Brainstorming in 2012. The 2nd Policy Support Brainstorming on "Safe Water and Sanitation in the New Millennium" was organized by NASI in collaboration with SISSO in Allahabad, September 2012. As in the 1st Brainstorming, the leading scientists, technocrats and sector leaders participated, and a number of papers dealing with important technical and policy issues were presented. At the end of the brainstorming, a set of recommendations were formulated which was submitted to key government departments and important R&D and HRD organizations. However, it was felt and Dr. Sharma strongly endorsed the idea that the papers presented in the conference contained important scientific, technical and policy issues which would be extremely valuable for the scientists and professionals of the country, and as such we agreed, and it was proposed to the NASI Council that a book containing these papers should be published. The proposal was approved by the NASI Council, and the internationally reputed publishing house Springer Nature has been entrusted with the publication of the book.

The book contains 20 chapters including the introductory overview of the country situation by the undersigned in Chap. 1. Chapter 2 deals with the present status of rural water supply in the country in the context of the current initiatives by the government. The sanitation scenario along with the flagship programmes including problems, prospects and key constraints is elaborately discussed in Chaps. 3 and 4. The concept of water safety plan, as visualized by WHO, and the guidelines for its applicability for the rural water supply projects in the country are discussed in Chap. 5. Various kinds of technologies applicable for the rural communities and the risk and safety requirement associated with the same, are thoroughly discussed in Chap. 6 including issues related to household point-of-use treatment and its health significance along with consumer safety. The geogenic and anthropogenic contamination of groundwater with arsenic, fluoride, pesticides, heavy metals, etc. and its impact on community health are elaborately discussed in Chaps. 7, 8 and 9. Issues related to arsenic-related technologies and case studies for community-based approach for the same are discussed in Chaps. 10 and 11. Urban wastewater management with particular reference to sewer rehabilitation and maintenance is discussed with elaborate illustrations in Chap. 12. The critical issues of urban solid waste management and the challenges posed by the same for the municipal managers are analysed and discussed in Chap. 13. The bioenvironmental control of vectors and the need for an integrated and eco-based approach are presented in Chaps. 14 and 15. The impact of bacteriological and chemical contamination of drinking water on the health and nutritional status of the community is highlighted in Chaps. 16 and 17. The sustainable management of water resources in India is discussed with particular reference to the feasibility of linking the rivers in Chap. 18. A most interesting case study on the cost-benefit analysis of production of hydropower from the river is critically discussed in Chap.19. In Chap. 20, the author discusses a novel Gandhian model of development based on non-violent culture.

Kolkata, India K.J. Nath

Contents

About the Editors

Prof. K.J. Nath, BE (Cal), MEPH (Cal), DSE (Delft). FIE, FNASc, FIPHE, FIWWA, FNESA is widely regarded as one of India's foremost experts on Sanitation and Community Water Supply, Environmental Health and Public Health Engineering. As Professor of Environmental Sanitation and Director of the All India Institute of Hygiene & Pubic Health (AIIH&PH), Govt. of India (GOI), he did seminal works on rural and urban sanitation and water supply, urban waste management, air and water pollution, water quality surveillance, environmental epidemiology and impact assessment. His expertise has been requisitioned by several National & International Organizations including the Planning Commission, Council of Scientific & Industrial Research (CSIR), Indian Council of Medical Research Institutes (ICMR). Ministries of Environment, Health, Urban Development, and Rural Development of the Govt. of India as well as State Govts. & Local and Development agencies, Research Academic Institutes. At the International level, he has served on a number of occasions as a consultant and Member/Chairman of expert panels of the World Bank, Asian Development Bank, WHO, UNICEF, UNEP, UNDP, WSSCC, UN-Habitat, International Scientific Forum on Home Hygiene (IFH), U.K, U.S Trade Development Agency etc. He served as a Member of Steering Committee/Working Group of the Planning Commission from the sixth 5 Year Plan to 10th 5 Year Plan. He was a Member of the Planning Commission High Power Committees on SWM and Public Health (1997–1998), Steering Committees of the 1st and 2nd Phase Ganga Action Plan, Expert Member of NGRBA (2009–2014), National Task Force on Environmental Health, Chairman of the National Co-ordination Council on Water Quality, Rajiv Gandhi Drinking Water Mission. Presently he is the **Chairman, Arsenic Task Force, West Bengal; Chairman, Core Committee on Water Quality, Safety & Security, Govt. of West Bengal and Member, Scientific Advisory Board and SEA Regional Coordinator, International Scientific Forum on Home Hygiene (IFH), UK, Member of the Technology Advisory Group, WHO (Climate Resilient WASH), President of Institution of Public Health Engineers, India (IPHE),** the premier national professional association of the Public Health Engineers in India. He has published more than 150 research and review papers, books, monograms etc. He has received several awards including Rajiv Gandhi Memorial Sulabh Sanitation Award (1998), Environmental Scientist of the Country award by National Environmental Science Academy (1999), Golden Jubilee award of the Indian Public Health Association (2007), Most Eminent Public Health Engineer of the Country by Institute of Engineers in 2008. **He is a fellow of the National Academy of Sciences, India; Indian of Public Health Association; Institution of Public Health Engineers, India; Institute of Engineers; Indian Water Works Association and he is an Honorary Fellow of the National Environmental Science Academy, India and Member, International Water Academy, Oslo. Prof. Nath has to his credit, more than 150 publications of books, monograms, course manuals, etc. and papers presented in International and National Conferences and published in reputed International and National Journals.**

Dr. Vinod Prakash Sharma, MSc (1960), DPhil (1964), DSc (1979), all from Allahabad University, is postdoctoral research associate of the University of Notre Dame and Purdue University, USA (1964–1968); pool officer of the Forest Research Institute, Dehradun (1969–1970); senior scientist of WHO/ICMR Research Unit on the Genetic Control of Mosquitoes, New Delhi (1970–1975); deputy director of Vector Control Research Centre/Malaria Research Centre (1975–1982); founder director of the National Institute of Malaria Research (1962–1998); and additional director general of the Indian Council of Medical Research (1998). He is FNA, FASc, FNASc, FAMS, FTWAS and fellow of several other learned societies. He has specialized in malaria and vector biology. He has 40 years of research and field experience, published 400 scientific papers and book on *Anopheles* and edited 14 books. Under his leadership, NIMR had earned international reputation of excellence in malaria research and control and established linkages with leading national and international laboratories. He conceptualized the bioenvironmental malaria control strategy as an alternative to spraying and successfully demonstrated malaria control in various eco-epidemiological settings. Dr. Sharma's work has become legendry. As a result of his indefatigable efforts, today we have technologies to fight malaria which are safe and cost-effective, produce sustainable impact and are free from environmental hazards produced by DDT and other insecticides. His work has been recognized nationally and internationally. He was invited to work on various WHO panels since 1980s. He was chair of the WHO/UNEP/FAO Panel of Experts on Environmental Management (PEEM) for Vector Control (two 5-year terms). WHO invited him to write/review GFATM proposals for SEA Regional countries, i.e. Indonesia, Timor Leste, Myanmar, Bhutan, Bangladesh and India, and for Sierra Leone by the NGO BRAC. He was the principal architect for writing World Bank and other projects for the Indian programme on malaria control. He chaired review committees of the NMEP, Nepal, Bhutan and Thailand, and is actively involved in the evaluation of countries capabilities to proceed towards malaria elimination. He was invited in WHO (Geneva and New Delhi) meetings as temporary adviser. He represented India on the Scientific Committee on the Problems of the Environment (SCOPE), Paris, for nearly 20 years and later was elected as executive member of the SCOPE. In that capacity he coordinated the "Health and Environment" cluster of the SCOPE. He was president of the National Academy of Sciences, India (1999–2000); Indian Society for Parasitology (1993–1997); and National Academy of Vector and Vector Borne Diseases (1979–2008); chief editor of the *Journal of Parasitic Diseases* and *Indian Journal of Malariology*; council member of the International Congress of Entomology, UK (two terms); and member of the WHO Expert Committee on Malaria, Geneva, still continuing. He is a recipient of several national and international awards/honours, e.g. WHO Darling Foundation Prize (1999); ICMR Ambedkar and MOT Iyengar awards; Lifetime Achievement Award; Green Scientist Award; G.P. Chatterjee Memorial Award; Ranbaxy award; FICCI award; Om Prakash Bhasin Award; Vasvik awards; distinguished parasitologist recognition by the World Federation of Parasitologists (2010); Gujar Mal Modi Award (2013); Vigyan Vibhuti award by the Uttrakhand State Council for Science and Technology (2013); Meghnad Saha Distinguished Fellow (2005–2010); NASI Distinguished Professor and ICMR Chair in Public Health Research (2010–2015) at the Indian Institute of Technology, New Delhi; and Padma Shri (1992) and Padma Bhushan (2014) awarded by the President of India.

Overview of the Current Scenario of Community Water Supply and Environmental Sanitation in India

1

K.J. Nath

1.1 Introduction

One of the greatest failures of the last 50 years has been the failure to lay the foundation stones of public health in the developing world – environmental sanitation and water supply. It is a failure that today deprives hundreds of millions not only of health but of productivity. It is a failure that undermines the normal mental and physical growth of rising generations. It is a failure that pollutes fresh water resources with faecal matter and organic pollutants on a massive scale. It is a failure that condemns more than a billion people to live with a daily environmental crisis of squalor, smells and diseases. And it is a failure that holds back the development of people and of nations. Despite significant progress made in the economic and industrial development, the demographic and environmental health scenario in the country continues to be a cause of serious concern. Water and air pollution along with malnutrition and lack of sanitation is increasing the disease burden in the society. The existing health infrastructure is unable to cope with the same.

It is a paradox that while India has progressed significantly in science and technology and is among the leading economic powers in the world, a substantial portion of the population does not have access to sanitation and safe water.

Chairman, Arsenic Task Force, Govt. of West Bengal; Member, Scientific Advisory Board and SEA Regional Coordinator, International Scientific Forum on Home Hygiene (IFH), Geneva; Chairman, Science and Technology, Sulabh International Social Service Organization; Vice Chancellor, International Academy of Environmental Sanitation and Public Health (IAES&PH); Former Director, All India Institute of Hygiene and Public Health, Govt. of India; Expert Member, National Ganga River Basin Authority (2009–2014), Govt. of India. Email: nathkj@hotmail.com

K.J. Nath (✉)
Institution of Public Health Engineers (India),
All India Institute of Hygiene & Public Health (GOI),
IPHE Bhawan, CK-58, Sector-II, Salt Lake City, Kolkatta 700091, West Bengal, India
e-mail: nathkj2016@gmail.com

K.J. Nath, V.P. Sharma (eds.), *Water and Sanitation in the New Millennium*,
DOI 10.1007/978-81-322-3745-7_1

India is now in the front ranks of fast growing emerging economies, it is also one of the countries wherein a lot of efforts are still required to eliminate the practice of open defecation.

In urban areas, faecal contamination of community water supply systems is continuing unabated. Pollution of Ganga and other national rivers is a serious threat to their survival. The investment in water supply and sanitation sector was negligible (less than 1%), during the first five 5-year plans. However, during the sixth plan which coincided with the launching of the International Decade of Water Supply and Sanitation, the allocation for water supply and sanitation sector was increased significantly (more than 4%). In recent years, the government has launched several flagship programmes like National Mission for Clean Ganga, Jawaharlal Nehru National Urban Renewal Mission and National Rural Drinking Water Quality Monitoring and Surveillance Programme. A significant development has been the launching of Swachh Bharat and Clean Ganga missions by the government of India in 2014. However, progress of implementation of these programmes and adapted methodologies needs serious review and policy change.

1.2 Looming Water Crisis

Water resource management and providing water security to the millions who are without the same is possibly the most challenging issue mankind would face in this century. The demand for water has increased significantly – by 700% during the last century. Almost 1800 million people globally will face serious water scarcity by 2015, and two thirds of the world population could be under stress. According to UN statistics, 1.2 billion people lack access to improved and safe water supplies, and if we consider microbiological and chemical contamination of water, this figure would be many times higher.

In 1947, the per capita water resources availability in India was measured at 5,150 cubic metres. By the year 2000, it decreased to about 2,200 cubic metres. It has been recently estimated that by 2017, India will "water stressed", and per capita availability will decline to 1600 cubic metres. Rapid population growth in the country will result in a further decline in the per capita availability of fresh water. Studies undertaken show that the amount of available aggregate annual utilizable water in India, surface and ground together, would be at about 1,100 billion cubic metres. Other problem areas are the fast-growing urban centres, where water requirements are expected to double by 2025. The situation concerning industrial supplies is even more difficult to analyse. Agriculture, the largest consumer of water resources, will probably utilize more than 70% of available water by the year 2025 to support the increasing food demand in the country.

1.3 Lack of Safe Water and Environmental Sanitation: A Critical Threat to Public Health

The World Health Assembly 1978 in Alma Ata adapted four key strategies for attaining health for all. One of these key strategies was "Promoting healthy lifestyles and reducing risk factors to human health that arises from environmental, economic, social and behavioral causes". If the agenda has remained unfinished by a wide margin, the primary reason could be found in our failure to develop an enabling policy for promoting a hygienic environment conducive to healthful living. The health outcome of the lack of sanitation and safe water is enormous – globally 4 billion cases of diarrhoea, 2.2 million deaths per annum and 62.5 million by disability-adjusted life years (DALYs) lost. The World Bank estimates that 99.9% of deaths attributed to poor water supply, sanitation and hygiene occur in the developing countries. To this, we must add the new menace of the chemical contaminations of groundwater sources, particularly arsenic and fluoride, not to forget about the problems created by salinity and iron and toxic heavy metals and pesticides. The gravity and enormity of the problem is at once evident when we consider that about 30% of our total usable fresh water resources of about 4500 billion m^3 is stored underground.

Air pollution is a serious problem in the urban areas which is causing acute respiratory infections, cardiovascular diseases, asthma, etc. This is also abating tuberculosis which continues to be a serious threat to public health because of poverty and malnutrition. Inadequate drainage and improper solid waste management are creating conditions conducive to vector breeding giving rise to vector-borne diseases like malaria, filaria, dengue, encephalitis, etc. Though a scientific and epidemiological assessment is yet to be made, the non-communicable diseases caused by factors related to lifestyle and environment could be serious threat to community health in the coming years.

1.4 Pollution of Fresh Water Sources and Declining Water Quality: A Critical Challenge

In the developing countries, major portion of the waste water (80% according to some estimate) is discharged into the water bodies without any treatment. This is primarily because of lack of resources with ULBs and also absence of appropriate regulation.

Calcutta was the third city in the world to have sewerage system in 1870 after London in 1850 and New York in 1860. Unfortunately, while the European and other developed nations saw a rapid growth of sewerage systems, in India, out of 5000 and odd cities and towns, only 232 are having sewerage systems that are too partially. Treatment of community waste water and industrial effluents discharging into major rivers and other inland water sources is extremely inadequate. In rural areas, open defaecation is as yet the major mode of sanitation, and this contributes significantly to non-profit sources of water pollution along with poor solid waste management. The water quality monitoring results obtained during 1995–2006 indicate that the organic and bacterial contamination are continued to be critical in water bodies. This is mainly due to discharge of domestic waste water mostly in untreated form from the urban centres of the country. The municipal corporations at large are not able to treat increasing the load of municipal sewage flowing into water bodies without treatment. Secondly the receiving water bodies also do not have adequate water for dilution. Therefore, the oxygen demand and bacterial pollution are increasing day by day. This is mainly responsible for waterborne diseases. *It requires serious introspection by the scientists and technocrats of the country whether the Western model of sewerage and waste water system is based suited to our socio-economic and demographic situations and urban structures. The on-site sanitation model and decentralized waste water system developed by Dr. Bindeshwar Pathak could be considered as an alternate approach.*

1.5 Urban Development and Ecological Crisis

Our approach to development and economic growth is not taking care of the ecological concerns. Under the pressure of escalating and indiscriminate urbanization and industrialization, rivers are vanishing into drains, and lakes and wetlands are evaporating from the urban maps. Just to mention, a few examples of many such cases in the country are the following:, Najafgarh drain, Delhi (Sahibi river), Tolly's Nullah (Adi Ganga), Buddha Nala (Ludhiana) and Mithi drain (Mumbai). There will be a time when our children will forget that the Yamuna, Cauvery, Damodar and Churni were rivers. I won't be far from truth if I say that we are committing hydrocide – deliberate murder of our water bodies, at the altar of urban greed.

Real estate and commercial malls are growing to whet the hunger of urban and commercial growth, and beautiful orchards and green gardens of yesterday are fast vanishing into the concrete jungle of high-rise buildings and shopping malls. In Hyderabad, the expanding concrete jungle is eating into the catchment of Himayat Sagar and Osman Sagar lakes – inflows dropping by 25–30%. The future of Kolkata wetlands is facing critical jeopardy. One wonders whether the growth crazy society is possessed by death wish! It needs serious introspection by the scientists, technocrats, planners and the policymakers whether the current model of economic development and growth is ecologically sustainable.

1.6 Restoration of "Nirmal Aviral Dhara" in Ganga

Former Prime Minister of India, Shri Rajiv Gandhi, launched Ganga Action Plan in 1987. Because of many reasons, the activities undertaken during the first phase and second phase of Ganga Action Plan failed to achieve the desired objectives of the programme, and Ganga remains unfit for bathing in its entire stretch from Gangotri to Gangasagar. Almost all other major rivers and their tributaries are also seriously polluted. A major policy thrust in this regard has been the formation of National

Ganga River Basin Authority. The Ganga River Basin Management Plan conceptualized and developed by the IIT Consortium envisages that the restoration of "Nirmal Dhara" in all rivers of the Ganga basin will require, among other actions, the following steps concerning sewage collection, treatment and disposal processes. Complete stoppage of the discharge of sewage, either treated or untreated, from towns into all rivers of the Ganga basin:

- All sewage must be collected and treated up to tertiary level (treatment guidelines for tertiary treatment specified elsewhere; effluent standards: BOD<10 mg/L, SS<5 mg/L, fully nitrated effluent, $P < 0.5$ mg/L, FC<100/100 mL).
- The treated water should be recycled or reused for various purposes, i.e. industrial, irrigation, horticulture, non-contact/non-potable domestic uses, groundwater recharge, etc.

It needs to be emphasized that if the targets and policies of the Ganga Authority are to be implemented effectively by the state government and adequately maintained by the local government authorities, a major capacity building programme has to be undertaken in all urban and rural local government authorities. Maintenance of minimum ecological flows (Aviral Dhara) in our rivers is a critical challenge, as it often clashes with our urban, industrial and energy development policies. Implementation of a number of hydroelectric projects has resulted in significant alteration in hourly, daily and seasonal flows over substantial river length, in the upper Ganga segment (Gomukh to Hardware).

1.7 Water Quality Surveillance in Rural and Urban Areas

Water quality monitoring and surveillance is not adequately taken care of in most of our urban local bodies. In most rural systems, it was nonexistent till recently. The most glaring example of nonexistent water quality surveillance system in the rural areas is the episode of arsenic contamination of groundwater sources in India and Bangladesh. People were using arsenic-contaminated sources for years, without the quality of water of sources, being tested even once. It is only after some of them got sick that the problem was identified. Unfortunately, the utility and importance of this vital function in the overall management of drinking water supply system is often not realized by the concerned authorities and beneficiaries as well. Administrative limitations, lack of government and political support and, on top of this, general public apathy are the primary factors behind the low priority so far accorded to these essential functions. To develop more effective operation and maintenance of urban and rural water supplies, it is required to promote sanitary inspection along with effective consumer level water quality monitoring and surveillance in the rural areas, as a mechanism to identify problems and to take corrective measures at the community level. In absence of adequate operation and maintenance of the distribution systems, lack of leak detection and preventive maintenances, an intermittent system of discontinuous supply results in almost universal faecal contamination of the distribution system and frequent outbreak of waterborne infections particularly in the unserved areas of the vulnerable groups (urban slums/peri-urban areas). In a recent water quality survey by the National Environmental Engineering Research Institute (NEERI) in all the major cities of towns in the country, it was found that 20–100% of the samples were faecally contaminated. In another survey in Kolkata, it was revealed that faecal contaminations in most of the household reservoirs were even more than in municipal systems. Achieving an acceptable standard of operation and maintenance and introduction of 24x7 supply systems would raise the question of affordability. However, it must be noted that urban India cannot enter the twenty-first century without introduction of continuous uninterrupted water supply to its citizens and protect them from microbial contamination.

1.8 Sanitation: The Unfinished Agenda

The International Decade on Water Supply and Sanitation was launched in 1980, and much was expected in terms of health benefit from the projects undertaken during the decade. Unfortunately, most of the national governments in the developing world provided an extremely disproportionate priority to sanitation in comparison to water supply. The propensity to give priority to water supply over sanitation has cost us hugely in terms of health gains. It is improved hygiene and sanitation – keeping faecal matter away from hands and food and from water itself when it is stored in the home – that transforms health. And the neglect of hygiene and sanitation goes a long way towards explaining why water supply programmes have often not brought the expected health benefits. The supply-driven, subsidized government programme did not succeed during the decade, primarily because lack of enabling environment (hygiene behaviour change + demand for sanitation), lack of capacity development at the community level/ULBs/PRIs, lack of IEC support and awareness among the consumers and sanitation were not a perceived need of the community. *The decade was a spectacular failure.*

Total Sanitation Campaign (TSC), launched in India during the 1990s, and community-led total sanitation (CLTS), launched in many developing countries, created the much needed enabling environment, triggered behaviour change in the community and generated demand for sanitation. But the improvement was not sustainable without adequate government support for a sustainable technology. Often toilets with flimsy construction materials and unsustainable technology were washed away and many Nirmal Grams, open defaecation-free (ODF) villages reversed to OD villages. In the context of developing countries, in majority of the rural households, economic decision might not be in favour of investment on sanitation, and as such market alone cannot be the key to promotion of sanitation, and an element of subsidy, as such, cannot be avoided altogether.

1.9 Swachh Bharat Abhiyan: Road to Success

The prime minister has launched the Swachh Bharat Abhiyan with lot of passion and a sense of urgency. His passionate appeal has created positive impact on various sectors of the community. However, it must be realized that the programme as such is not fundamentally different from the Nirmal Bharat Abhiyan launched by the previous government, and the success of the same would largely depend on mobilization of resources and total participation of the people. A positive impact of the prime minister's appeal has been the response from the corporate sector in mobilizing the CSR funds for sanitation promotion. We must learn from our past mistakes in policy planning and management. CLTS/sanitation marketing and government-subsidized programme need not be mutually exclusive; rather future sanitation programme should draw upon the strength of CLTS/sanitation marketing and integrate them into a scaled-up nationwide programme supported by the government and promoted by the sector partners with provision of minimal subsidies required by the poor for having access to sustainable sanitation.

1.10 Epilogue: Promoting Water Safety and Sanitation Pays

Providing safe water and sanitation to the community is a basic precondition for improving health and alleviating poverty. A World Bank study says India is losing 4%–6% of GDP in damages related to public health, unsafe water, lack of sanitation and environmental pollution. In this logic, a much higher allocation on water supply, sanitation and public health sector (close to 4% of GDP) would be amply justified. Gains for the developing world, from investment on water and sanitation, could reach 15 US$ per capita per year.

Prof. K.J. Nath, BE (Cal), MEPH (Cal), DSE (Delft). FIE, FNASc, FIPHE, FIWWA, FNESA is widely regarded as one of India's foremost experts on Sanitation and Community Water Supply, Environmental Health and Public Health Engineering. As Professor of Environmental Sanitation and Director of the All India Institute of Hygiene & Pubic Health (AIIH&PH), Govt. of India (GOI), he did seminal works on rural and urban sanitation and water supply, urban waste management, air and water pollution, water quality surveillance, environmental epidemiology and impact assessment. His expertise has been requisitioned by several National & International Organizations including the Planning Commission, Council of Scientific & Industrial Research (CSIR), Indian Council of Medical Research Institutes (ICMR). Ministries of Environment, Health, Urban Development, and Rural Development of the Govt. of India as well as State Govts. & Local and Development agencies, Research Academic Institutes. At the International level, he has served on a number of occasions as a consultant and Member/Chairman of expert panels of the World Bank, Asian Development Bank, WHO, UNICEF, UNEP, UNDP, WSSCC, UN-Habitat, International Scientific Forum on Home Hygiene (IFH), U.K, U.S Trade Development Agency etc. He served as a Member of Steering Committee/Working

Group of the Planning Commission from the sixth 5 Year Plan to 10th 5 Year Plan. He was a Member of the Planning Commission High Power Committees on SWM and Public Health (1997–1998), Steering Committees of the 1st and 2nd Phase Ganga Action Plan, Expert Member of NGRBA (2009–2014), National Task Force on Environmental Health, Chairman of the National Co-ordination Council on Water Quality, Rajiv Gandhi Drinking Water Mission. Presently he is the **Chairman, Arsenic Task Force, West Bengal; Chairman, Core Committee on Water Quality, Safety & Security, Govt. of West Bengal and Member, Scientific Advisory Board and SEA Regional Coordinator, International Scientific Forum on Home Hygiene (IFH), UK, Member of the Technology Advisory Group, WHO (Climate Resilient WASH), President of Institution of Public Health Engineers, India (IPHE),** the premier national professional association of the Public Health Engineers in India. He has published more than 150 research and review papers, books, monograms etc. He has received several awards including Rajiv Gandhi Memorial Sulabh Sanitation Award (1998), Environmental Scientist of the Country award by National Environmental Science Academy (1999), Golden Jubilee award of the Indian Public Health Association (2007), Most Eminent Public Health Engineer of the Country by Institute of Engineers in 2008. **He is a fellow of the National Academy of Sciences, India; Indian of Public Health Association; Institution of Public Health Engineers, India; Institute of Engineers; Indian Water Works Association and he is an Honorary Fellow of the National Environmental Science Academy, India and Member, International Water Academy, Oslo. Prof. Nath has to his credit, more than 150 publications of books, monograms, course manuals, etc. and papers presented in International and National Conferences and published in reputed International and National Journals.**

National Scenario of Rural Water Supply: Problems and Prospects

2

Dinesh Chand

2.1 Introduction

Safe drinking water is the basic necessity for life. Supplying of drinking water in rural areas has always been one of the prime goals of the government. The government policies and programmes in rural water supply have also undergone a series of changes ever since Independence. In the initial years, emphasis was laid upon providing physical infrastructures like sanitary wells, handpumps, etc. Thereafter, one can see a transition from implementation of simple drinking water supply systems to a techno-sociological approach seeking active participation of the people. Serious steps to ensure sustainability of the systems were initiated in 1999 to institutionalise community participation in planning, implementation and operation and maintenance of rural drinking water supply schemes through *Sector Reform Project.* Principally, *Sector Reform* ushers a paradigm shift from "government-oriented supply-led approach" to "people-led demand-centric approach".

Water which in turn includes rural drinking water supply is a state subject and has been included in the Constitution of India among the subjects entrusted to Panchayati Raj Institutions (PRIs) by the states. To accelerate the coverage of problem villages with drinking water, the government of India introduced Accelerated Rural Water Supply Programme (ARWSP) in 1972–1973 to support states/UTs through financial and technical assistance to implement water supply schemes. Later on in 1986, the "National Drinking Water Mission (NDWM)" was launched to accelerate the pace of coverage and resolve rural drinking water problem. Subsequently, it was rechristened as "Rajiv Gandhi National Drinking Water Mission (RGNDWM)" in 1991. This programme aimed at covering of all rural habitations with population of 100 and above, specially the uncovered habitations and having water quality-affected sources, and to set up water quality monitoring and

D. Chand (✉)
Ministry of Drinking Water & Sanitation, New Delhi, India
e-mail: d.chand@nic.in

K.J. Nath, V.P. Sharma (eds.), *Water and Sanitation in the New Millennium*,
DOI 10.1007/978-81-322-3745-7_2

surveillance system following "catchment area approach". The Tenth Five-Year Plan and subsequent plans emphasised participatory approach wherein PRI would act as nodal institution for convergence of drinking water supply programmes at grass-roots level to ensure sustainability of systems. The strategies to achieve the objectives of RGNDWM were as summarised below:

(a) To accelerate coverage of the remaining not covered (NC), partially covered (PC) and slipped back habitations from fully covered (FC) to PC or NC categories, with safe drinking water systems
(b) To provide safe drinking water to all water quality problem habitations and to set up water quality monitoring and surveillance system
(c) To implement water supply systems with sustainable sources ensuring continued safe drinking water supply in all covered habitations

As of 1 April 2005, 96.13% of rural habitations were fully covered (FC), 3.55% were partially covered (PC) and 0.32% was not covered (NC) with drinking water facilities, through millions of handpumps and over one lakh piped water schemes. However, there were slippages of FC habitations into NC or PC due to lowering of groundwater table or failure of sources, systems outliving their lives, increase in population, etc. In order to address the major issues like sustainability, water availability and poor water quality, etc., ARWSP Guidelines have been revised w.e.f. 1 April 2009 and renamed it as "National Rural Drinking Water Programme (NRDWP)" focussing on the following areas:

- To move forward from coverage of habitation towards coverage of households with drinking water
- To move away from overdependence on single drinking water source to multiple drinking water sources, through conjunctive use of surface water, groundwater and rainwater harvesting to ensure sustainability and prevent occurrence of slip-back habitations
- To encourage water conservation practices and revival of traditional water bodies
- To converge various water conservation programmes at the village level so as to ensure household level drinking water security adopting water budgeting and preparing village water security plans
- To consciously move away from high-cost treatment technologies to tackle arsenic and fluoride contaminations and to develop alternative safe sources for arsenic-affected habitations and adopt dilution practices in these aquifers through rainwater harvesting techniques to tackle fluoride contamination
- To establish water testing laboratories, at the district and subdivision levels, and to develop the capability for preliminary water testing at gram panchayat level
- To encourage handing over of rural drinking water schemes management (RWSM) to PRIs and earmarking of assets, and 10% of NRDWP-allocated funds as incentive to the states for transfer of management to PRCs was introduced

Further, one of the six components of "Bharat Nirman", which envisages to building strong rural infrastructure in 4 years (2005–2006 to 2008–2009), was to provide drinking water supply through coverage of over all not covered, slipped back and water quality problem habitations under the drinking water programme.

2.2 Challenges in the Rural Drinking Water Sector

2.2.1 Coverage and Moving Up on the Ladder

As per Census 2011, 75.71% of Indian households have access to drinking water within the reach of 500 m. However, according to recent estimates of the "Ministry of Drinking Water and Sanitation", 72.27% of the rural population have access to drinking water. Out of 1.71 million rural habitations in the country as of 1 April 2016, 1.31 million habitations were FC, 0.33 million PC and 71,077 remained NC which includes water quality-affected and slip-back habitations. In fact, from the 1990s, there has been a considerable reporting of increase in coverage of rural water supply during the successive 5-year plans right from the 1990s.

However, the increase in annual investment in the WATSAN sector does not translate in proportion with increase in access to drinking water (Figs. 2.1 and 2.2). Thus, there is an urgent need to address the twin problem of sustainability of water sources and their quality. MoDWS has estimated a huge gap in financial resources for tackling problems of sustainability and quality of water sources.

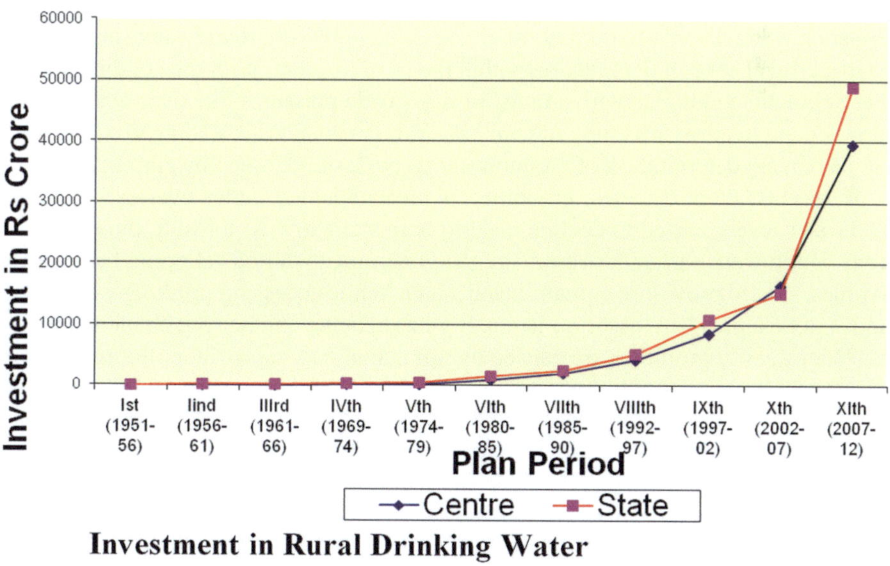

Fig. 2.1 Investment in rural drinking water

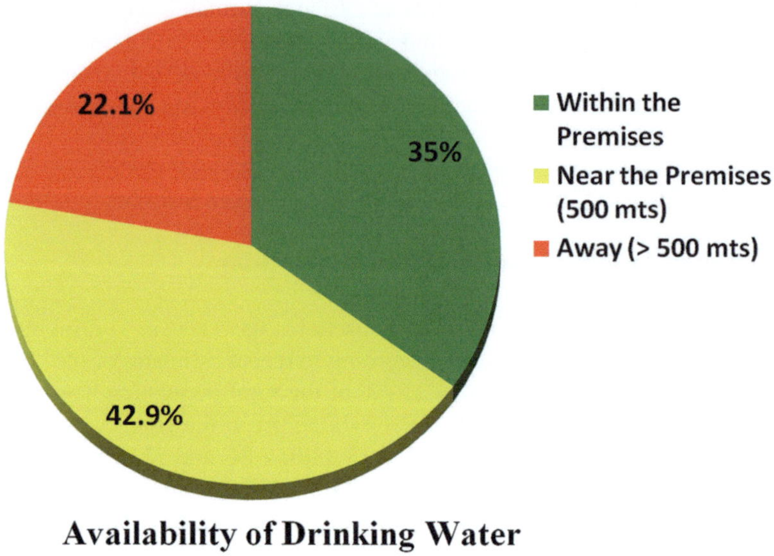

Fig. 2.2 Availability of drinking water

2.2.2 Sustainability

The average annual rainfall in the country is extremely abundant by global standards. However, much of this rain falls in relatively brief downpour during the monsoon, and there is great disparity across different regions. Groundwater is a critical water resource in India, as it accounts for 85% of drinking water sources and over 65% of irrigation water. However, current trend reveals that 60% of groundwater sources will be in a critical state of degradation within the next 20 years. With the combination of these climatic conditions with a range of man-made pressures like over-withdrawals of water and massive infrastructure building, this dependence is leading to a rapid and very worrying deterioration of groundwater resources underlined by current events.

In the last three decades, groundwater levels buckled under over-exploitation. The water levels started to decline making bore wells dry. As a result, those habitations which were earlier "covered" by the government started "slipping back" and, at times, started providing contaminated water. While accessing drinking water continues to be a problem; assuring its quality has become another big challenge.

Water quality problems are caused by municipal and industrial pollution as well as over-exploitation. With the rapid pace of industrialisation and greater emphasis on agricultural growth combined with financial and technological constraints, ineffective enforcement of pollution laws has contributed towards pollution of fresh water sources. These include sewage and industrial discharges and run-offs from agricultural fields and urban areas. Water quality has also got affected significantly by floods, droughts and due to lack of awareness and education among users. The need for users' involvement in maintaining water quality and looking at other aspects like safe water storage, hygiene, environment sanitation and proper disposal of waste and waste waters are critical elements of maintaining water quality.

2.2.2.1 Source Sustainability: Quantity and Quality

In places where surface water is available but unsafe for drinking or farming—more than 70% of India's surface water resources are polluted by municipal waste or toxic industrial chemicals—groundwater has generally been considered as a safe alternative. As far as the overall picture is concerned, presently, 178 districts (30%) have "unsafe" levels of groundwater development. Many of these districts also have serious water contaminations, but those districts which are considered "safe" in terms of availability of groundwater quantity need to be focussed. Among these "safe" districts, as many as 169 districts have one or more water quality problems (arsenic/fluoride or salinity). Thus, it was found that a total of 347 districts (about 59% of all districts in India) taking 169 districts having water contaminants together with another 178 districts with "unsafe" levels of groundwater development have problems of either inadequate availability or quality of groundwater. This clearly reveals that the coverage of most habitations in India has achieved from physical angle, but from water quality angle, it might be misleading.

Groundwater can get contaminated by a variety of sources including point and nonpoint sources of pollution and geogenic occurrences. As such pollution sources pose a relatively serious threat to a country's groundwater resources and need to be identified and tackled urgently. Point sources are often legally permitted to discharge a certain amount of substance and can be regulated by permitting procedures. However, there are also illicit pollutant discharges under this category from non-permitted sources which may cause serious water quality problems.

In contrast, nonpoint sources of pollution are not only hardly identifiable or locatable but also much complex to regulate as compared to point sources of pollution. Nonpoint source contaminants, naturally occurring substances, like arsenic, fluoride, iron, nitrates, salinity, other heavy metals, tannins, hardness and odour, often have unidentifiable source. However, when the polluters are unidentified, or the problem results from the behaviours and activities of residents, administration of control becomes more cumbersome.

2.2.2.2 System Sustainability: Quality of the System and Their Operation and Maintenance (O&M)

Over 5.69 million handpumps benefiting 620.6 million people and over 7.05 lakh piped water schemes benefiting 480.3 million people have been installed in the country so far under the Rural Drinking Water Programme. The total estimated cost for O&M of these as per the present provision (10–15% of the assessed capital cost/year) would be around Rs 2000 crore per year at present rate of funding. However, majority of the schemes remain non-functional, and many others become permanently defunct due to lack of skilled staff for proper maintenance and repairs and inadequate financial support. It is, therefore, necessary to give highest priority to O&M. Most of the states face resource constraints leading to poor maintenance. Funds under the state and NRDWP are provisioned to meet O&M costs which may not be adequate in most of the states.

In view of the huge cost of O&M, suitable institutional set-up and funding arrangements through community participation need to be evolved for upkeep of installations and working. The problem of poor maintenance can best be tackled by decentralising O&M by making the well-informed beneficiaries and panchayats responsible for O&M as already conceived in the sector reform programme. "Village Water and Sanitation Committees (VWSCs)", a subcommittee of panchayats, should actively be involved in the operation and maintenance of drinking water supply schemes. In this process, participation of village women and NGOs/voluntary organisations should also be encouraged. The funds available under NRDWP/Skill India Programme should be used to impart training to local youth, so that trained manpower can be deployed for the routine maintenance of the assets. Possibly, this trained manpower may be deployed under MNREGA/14th Finance Commission by PRIs for O&M activities. However, major repairs and replacement/rehabilitation projects may be allowed as plan schemes.

The steps were initiated way back in 1999 to institutionalise community participation in RWS sector by incorporating the following four basic principles in the reform project approach:

1. Adopting a demand-driven responsive approach based on empowerment of villagers to ensure their full participation in the project implementation through a decision-making role in the planning choice of scheme design and its implementation, control of finances and management arrangements
2. Empowering user groups (VWSC)/gram panchayats by the government for sustainable management of drinking water assets and integrated water conservation and management
3. Partial capital cost sharing either in cash or kind or both and 100% responsibility of O&M owns on community
4. If needed, greater for financial sustainability, convergence of other rural development programmes

Sector Reforms Projects were sanctioned in 67 districts on pilot based on the above principles. Later on, reform process has been scaled up after the experience gained from these pilot projects in the entire country through Swajaldhara launched on 25 December 2002. A notable feature of Swajaldhara is involvement of PRIs via formation of VWSC. Ten percent upfront contribution was made by the community, and 90% funds were provided by the central government. In case of SC and ST community, the contribution was in the form of cash, kind, labour or land or a combination of these. An integrated approach was adopted for water conservation aiming to augment the yield of water sources to provide sustainable drinking water supply to the rural population at 40 litres per person per day (lpcd).

2.2.3 Water Quality Management

Bacterial contamination in drinking water systems continues to be a widespread problem across the country and is a major cause of illness and deaths with 37.7 million people yearly affected by waterborne diseases. The major pathogenic organisms which cause waterborne diseases are bacteria (*E. coli*, *Shigella* and *V. cholerae*), viruses (hepatitis A, poliovirus, rotavirus) and parasites (*E. histolytica*, *Giardia*, hookworm).

WHO Guidelines for drinking water quality emphasise an integrated approach to water quality assessment and management from source to consumer stressing to be risk-based and quantitative. The guidelines also advise the maintenance of microbial quality to prevent waterborne infectious diseases as an essential goal. Further, these guidelines also stress upon water quality protection and prevention from chemical toxins and other contaminants of public health concern to be responsive to the underserved in developing countries which include of non-piped supplies and water collection, treatment and storage at household level to provide safe water. Beyond this there is a need for widespread use of an education and dissemination campaign that promotes and explains benefits of use of safe water among the people. Such a communication campaign is best done jointly by the different related sectors and stakeholders. Though use of safe water is one of essential needs for healthy living, adequate sanitation and proper nutrition interventions are also needed. All these three contribute to reducing diseases and increasing health, and the lack of one can adversely impact the benefits of the others. The importance of safe water, sanitation and nutrition to human health and well-being can be stated no better than UN Secretary General Kofi Annan in "Freedom from Want" in the Millennium Report, 3 April 2000, and reproduced as "How can we call human beings free and equal in dignity when over a billion of them are struggling to survive on less than one dollar a day, without safe drinking water, and when half of all humanity lacks adequate sanitation?"

The involvement of community in maintaining water quality and other aspects like hygiene practices, environmental sanitation, safe water storage and disposal of waste water are critical elements to maintain the quality of water resources.

In February 2006, the "National Rural Drinking Water Quality Monitoring and Surveillance Programme" under NRDWP was initiated by the government of India to institutionalise community participation in water quality monitoring and surveillance of drinking water sources at the grass-roots level by gram panchayats/VWSCs. However, their confirmatory tests for contaminations are yet to be conducted by district/state level laboratories.

2.2.4 Creation of Enabling Environment

2.2.4.1 Supporting Awareness Drives

One of the major challenges is to make people aware of handling and consumption of safe water. People drink water from contaminated surface sources despite being provided potable water systems by the government. The government needs to support the civil society and organisations to increase awareness. An integrated campaign can result in widespread information dissemination among the masses to prevent contamination of water sources.

2.2.4.2 Testing and Remedial Action

There is an urgent need to create the water quality monitoring network by establishing monitoring stations across all regions and seasonal assessments of all water sources. In case contamination is detected, a protocol for dealing with such sources should be available. The challenge lies in establishing well-equipped laboratories with adequate well-trained staff. This also calls for infrastructure development and capacity building of people. Although there has been wide usage of field testing kits, but they often give false or semi-quantitative results. One can rely on field testing kits only for identifying areas rapidly, but laboratory confirmatory tests are necessary for accurate results and correct actions. The data in respect of water quality affected habitations is available on the website of MDWS but many of the state water and sanitation departments do not display such data.

Further, collection of data, its interpretation and thus communication is an essential task for effective management of drinking water, and the use of geographical information system (GIS) may be useful in mapping, modelling and decision-making.

2.2.4.3 Institutionalisation of Decentralisation

Though planning for rural water supply is made at the central and state levels, responsibility for proper implementation has to be borne by the local bodies/PRIs. Under Article 243G of the Constitution, the state legislatures may, by law, endow the panchayats with the powers and authority necessary to enable them to function as institutions of self-government. Further, such law may contain provisions for the devolution of powers and responsibilities with respect to the preparation of plans, implementation of schemes for economic development and social justice and those relating to matters in the Tenth Schedule, which include drinking water and maintenance of community assets. So far only 23.64% gram panchayats have been given the responsibility.

As such, PRIs are the key institutions at the grass-roots level for drinking water supply programmes. However, the financial and administrative authority has not been devolved to them to the required extent. In this process, emphasis must be laid on the participation of stakeholders at various stages of project from planning, designing and location to implementation and management against presenting all responsibilities which are entrusted to line department. However, experience has shown that panchayats are unwilling to shoulder the responsibility of implementation of projects and their O&M due to non-existence of effective manpower and machinery at the village level for maintenance of the assets.

This calls for a radical change in the management system so that the decisions relating to implementing water supply schemes should be entrusted on local choices, demand and capabilities to meet the responsibility for O&M after taking the people into confidence and conducting consultations. PRIs should be made aware of the technologies, O&M costs involved giving the freedom to make their own choices, and simultaneously capacity building may be done. People's participation at all stages of the project implementation is likely to control the problem of sub-standard materials, poor workmanship and inadequate maintenance.

The participatory approach, which was a key element of the sector reform programme, should be seriously adopted. However, while part of the project costs should progressively be borne by the beneficiary community, the major source of funding for rural drinking water schemes may remain as the budgetary support from central and state governments till all rural habitations are provided with satisfactory and sustainable drinking water facilities.

2.2.4.4 Capacity Building of Communities

The roles and responsibilities of panchayats are becoming more important, and further stress is being laid on community-based approaches in dealing with drinking water-related problems. A prerequisite for increasing community participation is capacity building of the communities so that they are able take responsibility and make well-informed decisions. The objectives of decentralisation can be implemented successfully only if there is an attitudinal change among government functionaries as well as the people towards respect to decentralisation, transfer of authority and responsibility to the community. The role of the government in implementing capacity-building activities is essential. In Gujarat, village level water quality monitoring teams called *Gram Mitras* are being evolved for monitoring and strengthening water quality at the village level. Trainings are organised at the block level by government department for block level functionaries with the support of NGOs. As hands-on exercise, *Gram Mitras* collect samples from a drinking water source in the village and also conduct the sanitary survey of these sources. During the training, the testing of these sample links between the sanitary situation around the source and the quality of water is established. These *Gram Mitras* undertake water quality monitoring of water sources in villages across the state and also conduct awareness on ways and means to keep surroundings of drinking water sources clean.

2.2.4.5 Effective Groundwater Legislation and Enforcement

Recent data analysis reveals several alarming trends in the status of groundwater resources in India. The rate of withdrawal of groundwater is "unsafe" in 31% of the districts having 33% of the land area with 35% population. The situation has dramatically worsened within a short span of 16 years, between the assessments done in 1995 and 2011. Further, many of the so-called safe districts have severe problems of water quality, threatening their drinking water security. Taking quantitative and qualitative aspects together, it would appear that a total of 347 districts (59% of all districts in India) are vulnerable in terms of safe drinking water in India.

The groundwater safe level development in a district or block can be viewed from two angles, i.e. water quantity and its quality. MoDWS water quality data reveal that out of 593 districts for which data is available, the districts have quality problems of high fluoride (203 districts), iron (206 districts), salinity (137 districts), nitrate (109 districts) and arsenic (35 districts) (DDWS 2006). Biological contaminations which cause enteric disorders probably are a major problem present throughout the country linked with infant mortality, maternal health and thus loss of valuable "work time".

In India, however, it would seem almost impossible for the central government to manage the estimated over 25 million existing groundwater extraction structures as concerned institutions require significant strengthening. Moreover, the responsibility and functions of groundwater management are fragmented throughout the country among different government departments. In fact, state governments have primary jurisdiction over groundwater usage, and, in many cases, state agencies are poorly equipped. Both underground aquifers and above-ground rivers traverse the borders of states; competition over water use is already a major source of interstate conflict, as well as between users at a local level. To date, the difficulties of regulation and collective management of groundwater resources are a fundamental cause of the state of crisis. A World Bank study revealed that high-level policy reform in the shape of regulatory measures, economic instruments or tradable groundwater extraction rights is simply not a credible way forward and suggested that "bottom-up" community management may be the only way out. Other studies have supported this approach, with particular focus on community level groundwater recharge and the use of communally managed alternatives to groundwater, such as small dams.

There is no denying the fact that water is a basic necessity for the survival of mankind. There is interplay of various factors that govern access and utilisation of water resources. It becomes important to evolve holistic and people-centred approaches for drinking water management keeping in view the growing demand for drinking water.

2.2.4.6 Inter-agency Coordination

There is a fragmented approach at the state and central level among numerous agencies for the supply and management of water. One major bottleneck in an effective policy formulation and implementation has been lack of coordination among the stakeholders and the current institutional set-up involving various government agencies. Better coordination among ministries and departments would ensure effective implementation of the programme activities. The option of a single nodal ministry with the overall supervision and administration pertaining to water resources and their users may be considered in line of other countries like Australia.

2.2.4.7 Making the Service Provider Accountable

Under Article 21 of the Constitution of India, the Protection of Life and Personal Liberty and the right to pollution-free water are guaranteed under this provision. The user has the right to know whether water being provided at source is safe and free from any contamination as claimed by authorities. Financial expenditure on water supply schemes and data of water quality testing should be known to the public. The example of Tamil Nadu Water Supply and Drainage Board should be emulated by other states where financial expenditure is known to the public domain through their website.

2.3 Conclusion

Making provision of safe drinking water to all in rural India is a challenging task. Given the diversity of the country and its people, solutions have to be diverse and unique. One has to consider an approach that seeks the participation of users through interventions engaging the communities with various government policies and schemes. Such an integrated approach which includes planning, designing, implementing of water supply schemes and their operational management would incorporate collaborative efforts of various actors involving the government, civil society and needless to say, the people at large. Citizens should be made aware of the demand for clean and safe drinking water as a right as well as their responsibility through strong awareness programmes.

References

1. Annual Reports, Ministry of Drinking Water and Sanitation
2. Water Aid India (2005) Drinking water and sanitation status in India. Water Aid India, New Delhi
3. WHO (2011) Guidelines for drinking-water quality, 4th edn. WHO, Geneva
4. Ministry of Water Resources-website
5. GOI (2010) Guidelines for National Rural Drinking Water Programme (NRDWP), Ministry of Drinking Water and Sanitation
6. Strategic Plan – 2011–2022- Ensuring Drinking Water Security in Rural India, Ministry of Drinking Water and Sanitation, New Delhi
7. Sustainable Groundwater Management: Challenges – For the 21st Century
8. Kulkarni and Shankar (2009) India's groundwater challenge and the way forward
9. Paul Wyrwoll (2012, August 8) India's groundwater crisis: Solutions simply need to be found

Dr. Dinesh Chand is a professional engineer and has obtained his BEng (civil) from the University of Gorakhpur, MEng (environmental) from the Indian Institute of Technology (formerly University of Roorkee) and PhD from the University of Rajasthan.

He has over 36 years of experience in the field of water and air pollution control and water supply and sanitation sector related to sectoral planning and implementation aspects. During the above period, he held various posts starting his career as assistant environmental engineer (from May 1981 to September 1986) in the UP Pollution Control Board, assistant adviser and deputy adviser (October 1986–October 2008) in the Department of Drinking Water Supply (Rajiv Gandhi National Drinking Water Mission) (Government of India) and Joint Adviser (WR) (November 2008–February 2011) in the Planning Commission of India. He also worked briefly as Joint Adviser in CPHEEO in the Ministry of Urban Development from March to November 2011 before joining back as Additional Adviser (PHE) to the Ministry of Drinking Water and Sanitation (MDWS). He is also Chief Engineer of the International Centre for Drinking Water Quality (under MDWS), Kolkata. He has about 35 papers published in his credit and visited about 11 countries and also represented the Government of India in various ministries/departments, viz. meetings, reviews/ expert committees/teams, working groups/task forces, conferences, etc. and headed couple of expert teams/committees constituted by Govt.

Nirmal Grams: Problems and Prospects

3

Chandan Sengupta

3.1 Introduction

Quality of life is largely dependent on the availability of adequate safe drinking water and proper sanitation. Till the end of the twentieth century, availability of these two facilities in rural areas of India was miserable. The Rural Sanitation Programme (RSP), in particular, was never a priority in India. Until 1986, the RSP was looked after by the Urban Development Department. In that year, the program was entrusted with the responsibility of the Rural Development Department when the Central Rural Sanitation Programme (CRSP) was launched by the Government of India.

3.1.1 Central Rural Sanitation Programme

The CRSP was a subsidy-oriented program. Promotion of sanitation was considered as only provision of latrine at household level. Without creating any demand, latrine of better specifications was constructed at the villager's premises. Attempt for any contribution from the beneficiary was also not taken. Allocation for the program was also insignificant. Thus, few latrines provided through the program were not used by the villagers. During a survey, conducted in the late 1980s in West Bengal, it could be noticed that most of the latrines provided through the CRSP were used by the villagers for other purposes like storage of domestic animals, firewood, etc. All these factors resulted in very poor coverage of rural households with individual household latrines. This is why during the 1991 census, it was noticed that only 9.48% of rural households in India had access to individual household latrines. Poor achievement in coverage with household latrines through implementation of CRSP was also reflected in the census 2001. Access to household latrines in rural India increased to the extent of only 21.92%.

C. Sengupta (✉)
Former WASH Specialist, UNICEF, Kolkata
e-mail: csengupta124@yahoo.com

K.J. Nath, V.P. Sharma (eds.), *Water and Sanitation in the New Millennium*,
DOI 10.1007/978-81-322-3745-7_3

3.1.2 Total Sanitation Campaign (TSC)

In 1999, implementation of the whole sanitation program was critically reviewed. It was decided to expand the concept of sanitation by including personal hygiene, home sanitation, safe water, garbage disposal, excreta disposal, and wastewater management. Accordingly, in that year, the CRSP was converted into a campaign mode. The program was then named as Total Sanitation Campaign (TSC). The TSC adopted a demand-driven strategy where more emphasis was given on information, education, and communication (IEC), human resource development, and capacity development activities of all the stakeholders including nongovernment organizations at different levels. Stress was also given to increase awareness among the rural people and generation of demand for different kinds of sanitation facilities.

3.1.2.1 Alternate Delivery Mechanism

To facilitate implementation of sanitation as a package and also to give the villagers an option to choose cost-effective affordable model of sanitation facilities, it was decided to establish an alternate delivery mechanism after setting up production centers or sanitary marts. The main intention of this decentralized delivery mechanism was to run the program by the local NGOs or capable community-based organizations (CBOs) at village level. In other words, it was intended to implement the program with focus on community-led and people-centered initiatives. Financial incentive was allowed only for the below poverty line (BPL) households, not to the above poverty line (APL) households. Again, to encourage ownership and ensure use of the latrines, the amount of incentive was lowered substantially. Decision was also taken to provide toilets in all educational institutions and Anganwadi centers and at places of public gatherings (community sanitary complexes). To ensure overall environmental cleanliness in rural areas, safe disposal of solid and liquid waste management (SLWM) was also included within the TSC.

3.1.2.2 Nirmal Gram Puraskar (NGP)

In 2004, the Government of India introduced an innovative award in the name of *Nirmal Gram Puraskar* (NGP). The main objectives of introduction of this prestigious national award were:

(a) Achieve full sanitation coverage after providing individual latrines to all families, toilet blocks in schools and Anganwadi centers, and community toilet blocks in places of public gatherings in a defined panchayat area.
(b) Ensure use and proper maintenance of all toilets provided at household and institutional level so that open defecation-free (ODF) environment could be established within the Panchayati Raj Institution (PRI) body.
(c) Ensure safe disposal of solid and liquid wastes generated at households and at places of public gatherings so as to ensure clean environment within the villages.
(d) Create an environment of healthy competition among the implementers and also to recognize the achievements and efforts made in ensuring full sanitation coverage.

(e) Create an opportunity for a rapid scaling up of the TSC.
(f) Create an environment of great challenge to ensure that the spirit of the NGP was not diluted and the quality of the award was maintained.
(g) Encourage the panchayat bodies to take measures for sustenance of the program.

The award gained immense popularity and contributed effectively in bringing about a movement in the community for attaining the *Nirmal* status, thereby significantly adding to the achievements made for increasing the sanitation coverage in the rural areas of the country.

3.1.2.3 Achievement Through Implementation of TSC

In 2012, the program was again reviewed. It could be noticed that during 13 years of implementation of the TSC, commendable progress could be made. In this context, the status of the program in the country as of July 2012 (as reported in the portal of GOI) is mentioned below:

(a)	Access to IHHL:	Approximately 70.00%
(b)	Access to toilet in schools:	Approximately 90.00%
(c)	Access to toilet in Anganwadi centers:	Approximately 78.00%
(d)	No of NGP awarded GP:	28,002 out of total 2,38, 054 GPs (11.76%)

It seemed that the MDG for India in the sanitation sector was achievable.

3.1.3 Crisis Noticed in the TSC

Census 2011 revealed some uncomfortable facts about availability of sanitation facilities in the country. Instead of rosy pictures reported in the GOI website during that time as 66.66%, access to individual household latrines in rural India was reported 30.7% by the census.

3.1.3.1 Status of Availability of Sanitation Facilities in the Country as per Census 2011

The status of availability of household latrines in rural India as per online reports vis-à-vis census 2011 report of few major states has been shown in the table below:

S. No.	State	Access to toilets	
		As per online report (%)	As per census 2011 (%)
1	Tamil Nadu	83.63	23.2
2	Uttar Pradesh	78.15	21.8
3	West Bengal	81.37	46.7
4	Gujarat	82.99	33.0
5	Haryana	82.75	56.1
6	Maharashtra	65.88	38.0
7	Bihar	28.91	17.6
	India	**66.66**	**30.7**

A study conducted by the Joint Monitoring Committee of UNICEF and WHO on Sanitation during March 2012 also pointed out that 59% of the Indians still defecated in the open.

3.1.3.2 Status of NGP-Awarded GPs

To assess the spirit, principles and quality of the NGP status in the awarded Gram Panchayet (GP) was conducted by UNICEF in June 2008. The study covered 162 NGP-awarded GPs across 6 study states of the country (Andhra Pradesh, Chhattisgarh, Maharashtra, Tamil Nadu, Uttar Pradesh, and West Bengal). The study revealed a very sad state of affairs in the NGP-awarded GPs.

3.1.3.3 The Key Findings of the Study

Key findings of the study are mentioned below:

- Around only 81% of the households had access to household latrines.
- In 26% of the GPs, hardly 50% of the households had access to household latrines, and in 10% of the GPs, less than 50% households had access to household latrines.
- 60% of the households reported using their latrines, 5% of the households used community latrines, and 35% of the households went for open defecation.
- In families where the latrines were used, all the family members were not using their latrines.
- There was lot of signs of open defecation in the GP areas.
- Clean environment was not visible in the awarded GP area.

Toilets in NGP awarded GP

ICDS toilet not in use

3.1.3.4 Reasons for Nonuse of Latrines

The main reasons that could be understood for nonuse of latrines during the study are:

- Poor and/or unfinished installation of latrines (31%)
- Lack of behavioral change (17%)
- Absence of superstructure (14%)
- Blockage and/or chocked pan (12%)
- Lack of water (9%)

It was found that around 45% of the households disposed child's feces in the open or along with their solid wastes or in the drain.

3.2 Nirmal Bharat Abhiyan (NBA)

Bare facts revealed by the census 2011 on real situation of the sanitation scenario in India prompted the policy makers to revamp the whole program. It was thought that without any radical change in program strategy, it would never be possible for India to achieve MDG. Accordingly, in July 2012, the Government of India launched NBA in the country. The main objectives of NBA of course remained almost the same as that was for TSC. However, an attempt was made in the NBA to address the shortcomings of the TSC and also to accelerate the sanitation coverage in the rural areas.

3.2.1 Opportunities Available to the Implementers Due to Launching of NBA

The opportunities available in the NBA are mentioned below:

(a) Stress has been given on ensuring universal coverage of both BPL and identified APL households with individual household latrines. Provision of incentive has been extended to some identified disadvantageous categories of APL families belonging to:
   - SCs and STs
   - Small and marginal farmers
   - Landless laborers with homestead
   - Physically handicapped
   - Women-headed households

(b) Stress has been given on provision of individual household latrines of improved specifications at a cost of at least Rs. 5500.00. Accordingly, the amount of incentive to the BPL families was increased to Rs. 4600.00 per family. The balance amount is to be contributed by the beneficiaries. In case of difficult areas, the incentive has been increased to Rs. 5100.00, and the latrine could be provided at a cost of Rs. 6000.00.

(c) Appropriate convergence with MGNREGS with unskilled man-days and skilled man-days. In such case, additional amount of Rs. 4500.00 could be arranged for construction of latrines.

(d) Stress has been given on extensive capacity building of the stakeholders like PRI bodies at different levels, Village Water and Sanitation Committees, SHGs, and other field level functionaries.

(e) The states have also been advised to reassess the situation of accessibility of household latrines critically after carrying out GP-based detailed baseline survey to have clear ideas about (i) households with functional toilets, (ii) households with defunct toilets, and (iii) households belonging to both APL and BPL families having no toilets so that the NBA project could be formulated afresh for each the districts.

(f) Stress has been given on accelerating sanitation coverage in rural areas to achieve the vision of Nirmal Bharat by 2022 with all gram panchayats in the country attaining *Nirmal* status.

3.2.2 Strategy Suggested to Implement the Program to Attain the Status of Nirmal Bharat

Due to launching of NBA, a lot of opportunities are now available to the implementers. The program now is to be implemented with more care, and the errors that have been committed earlier are to be taken care of. Suggested strategies are mentioned below:

(a) The program is to be implemented as a demand-driven one. Thus, more stress is to be given on IEC activities. Funds are also now available for this purpose. Till now, hardly 30 % of the fund allocated for IEC has been spent by the states. A comprehensive IEC strategy is to be formulated, giving stress on interpersonal contact through trained SHG members. An area-specific IEC strategy is also to be launched.
(b) Stress is to be given on provision of individual household latrines of better quality. Area-specific innovative technology using locally available materials and labor is to be used.
(c) Mahatma Gandhi National Rural Employment Guarantee Act (MGNREGA) is to be successfully integrated with the NBA.
(d) Sustainable delivery mechanism is to be developed. Attempt is to be made to involve the local SHGs in the delivery mechanism.
(e) Stress is to be given on sustenance of the NGP (Nirmal) status involving local PRI bodies, grass root level functionaries, and the SHGs.
(f) The delivery mechanism is to be continued even after achieving the NGP status. Sustenance of the delivery mechanism is very important for repair, maintenance, and upgrading of toilets even after achieving the NGP status.
(g) Provision of running water supply for school sanitation is very important. In most of the schools, two toilet blocks are provided (one for boys and the other for girl students). In addition to this, midday meals are also provided in the schools. Provision of adequate running water supply is, thus, not only very important for maintenance of the toilets, it is also a must for promotion of health and hygiene education among the students.
(h) The online reporting system has been found to be not dependable. In the past, it has been noticed that there is a tendency for overreporting. Thus, there is a need for surveillance on the online reporting system.
(i) Stress is to be given on capacity development of local women groups including SHGs.
(j) Stress is to be given on empowering the PRI bodies at different levels so that they could plan, implement, and develop self-sustained program.
(k) In most of the states, almost no work has been done in the area of SLWM. Stress is to be given to develop GP-based SLWM schemes. Wherever possible, community-based schemes could be initiated. However, in our country most of the households are located in a scattered manner where it is not possible to promote community-based schemes. In such areas, SLWM is to be promoted at household level, giving stress on vermicomposting. Each state should develop area-specific guidelines integrating MGNREGAS for this purpose.
(l) Attempt is to be taken to enhance political will at different levels.

3.3 Conclusions

Launching of NBA has created huge opportunity to the implementers. Capturing the opportunities, the implementers can boost implementation of the sanitation program in the country. The policy makers has given all the implementing states a scope to review the program for making good the errors committed earlier. If the program is now implemented with all sincerity after capturing the available opportunities, there should not be any difficulty to achieve the vision of *Nirmal* Bharat by 2022 with all gram panchayats attaining *Nirmal* status in the country.

Acknowledgments (a) UNICEF office for Tamil Nadu, India; (b) India NGO, Chennai, Tamil Nadu; (c) Panchayat and Rural Development Department, Government of West Bengal

References

Guidelines issued by the Government of India on "Nirmal Bharat Abhiyan":

(a) Study conducted by the Women Coordinating Council, Kolkata on implementation of CRSP for the PHED, Government of West Bengal.
(b) Online progress report of the TSC and NBA as reflected in the website maintained by the Ministry of Drinking Water Supply, Government of India.
(c) Census 1991, 2001 and 2011 data published by the Registrar General of Census, Government of India.
(d) Study conducted by the TARU on status of Nirmal Gram Puraskar awarded GPs in India on behalf of UNICEF, India.
(e) Study conducted by the author on status of Nirmal Gram Puraskar awarded GPs in Tamil Nadu. Photographs attached were taken in NGP awarded GPs of Tamil Nadu.

Mr. Chandan Sengupta is an engineer by profession. He obtained B.E. (Civil) from Bengal Engineering College, Shibpur, West Bengal and M.E (PH) from the All India Institute of Hygiene and Public Health, Kolkata. He had undergone courses on Project Planning, Implementation and Appraisal of Water and Sanitation Projects in the Bradford University, U.K. and; on School Sanitation and Hygiene Education in the I. R. C., Delft. He has also been extensively trained in the area of Social Marketing, Participatory Rapid Rural Appraisal, Participatory Monitoring and Evaluation.

Mr. Sengupta is closely involved in the Drinking Water and Sanitation sector for more than last 35 years. He served the Govt. Of Tripura for about 22 years and was in charge of Water and Sanitation Programme in the whole state while he left to join UNICEF in the year 1989. He served the UNICEF for about 16 years and was in charge of Water and Environmental Sanitation Section in the UNICEF Kolkata office. After retirement from UNICEF, he served the Govt. Of West Bengal for more than 7 years as Chairman, Task Force on TSC. At present he is a freelance professional and delivers lectures in different states and engineering institutions. He is also member of the Scheme Selection Committees of Water and Sanitation Programme, Govt. of Tripura.

He is also a bird watcher and completed a professional course on Ornithology in the Institute of Bird Studies and Natural History, Rishi Valley Education Centre, Andhra Paradesh.

Mr. Sengupta is Life Fellow of IPHE, India and Indian Water Works Association (IWWA).

4 Promoting Rural Sanitation: Key Challenges

A.K. Sengupta

4.1 Introduction

In rural India, people had been practising open defecation for centuries. Sanitation, especially in rural areas of the country, never got priority in the developmental planning, and hence the census 2001 showed a dismal figure of hardly 20% coverage of sanitation in the country. Another factor that mattered much was low awareness of hygiene behaviour amongst the masses. This coupled with low sanitation coverage posed a very big challenge for rural India in the present century.

One can only lead a healthy life if he has proper water supply of adequate quantity and potable quality and proper sanitation. There is, therefore, a direct relationship between water, sanitation and health. In majority of developing countries, there is disproportionate disease burden due to the consumption of unsafe drinking water, improper disposal of human waste, inadequate handling of liquid and solid waste and lack of knowledge about personal hygiene and food hygiene. India is no exception to these facts, and high infant mortality rate has been indirectly attributed to poor sanitation practices in this country. All these issues led to the launching of the Central Rural Sanitation Programme (CRSP) in the country mainly to enhance the quality of life and provide privacy and dignity to the women folk and better facility to children and elderly.

In the early days, the concept of sanitation meant disposal of human excreta by any means such as cesspools, open ditches, pits, bucket system, etc., but the present concept of sanitation covers the whole gambit of environmental sanitation including solid and liquid waste management; personal, domestic and environmental hygiene; and food hygiene apart from human excreta disposal as well as potable water supply. In fact proper environmental sanitation is public health issue and is vital for not only

A.K. Sengupta (✉)
International Academy of Environmental Sanitation & Public Health,
Palam-Dabri Road, New Delhi 110045, India
e-mail: ak.sengupta48@gmail.com

K.J. Nath, V.P. Sharma (eds.), *Water and Sanitation in the New Millennium*,
DOI 10.1007/978-81-322-3745-7_4

individual but community health and social life. Environmental sanitation is one of the basic determinants of health, quality of life and human developmental index.

The responsibility for provision of sanitation in the country mainly vests with the local government functionaries – District Panchayat, Block Panchayat and Gram Panchayat in rural India. The central and state governments act as facilitators by providing policy, financial and capacity-building supports. Planning Commission provides guidelines for sector allocation funding and determines strategic priorities.

4.1.1 Global Scenario and Joint Monitoring Programme (JMP) 2012

JMP is Joint Monitoring Programme for Water Supply and Sanitation initiated jointly by WHO and UNICEF. This is being done every 2 years. The JMP 2012 report says, "Progress in China and India is highlighted, since these two countries represent such a large proportion of their regional populations. While China has contributed to more than 95% of the progress in Eastern Asia, the same is not true for India in Southern Asia. Together, China and India contributed just under half of the global progress towards the MDG target in sanitation. 11 countries make up more than three quarters (76%) of the global population without improved sanitation facilities. One third of the 2.5 billion people without improved sanitation live in India".

Nine hundred forty-nine million people as per the JMP 2012 report practise open defecation. Majority of them live in the rural areas. Open defecation in rural areas is being practised in almost all the regions of the developing world. Nine percent of Northern Africa and 17% of rural dwellers in Latin America and the Caribbean practise open defecation. But the highest percentage of the rural population, around 55%, who live in Southern Asia, are still practising open defecation.

The JMP 2012 has also done a comparative study of progress in urban and rural sanitation. The study shows that globally 79% of the urban population are practising improved sanitation, whereas hardly 47% of the rural population have improved sanitation facilities. Globally 1.8 billion people lack improved sanitation facilities, which is around 72% of unserved people. However, since 1990 sanitation facilities have been provided to more than 724 million of the rural population as reported in the JMP 2012 report.

4.1.2 Government Initiatives

The Central Rural Sanitation Programme was initiated by the Ministry of Rural Development, Government of India, in the year 1986, focusing mainly on providing household toilet to the rural community. The biggest missing link was the concept of behaviour change with the help of communication practices, and it did not advocate the total sanitation concept of personal hygiene, solid and liquid waste management and school and institutional sanitation. This approach totally failed to motivate and sustain the high level of sanitation coverage. In spite of an investment of more than Rs. 600 crores, rural sanitation grew at a very low rate of 1% per annum, and census 2001 found that hardly 22% of the rural population have been provided with sanitation facilities. This called for change in approach, and by 1999, the Government of India revised the programme and named it "Total Sanitation Campaign (TSC)".

TSC's main focus is participatory and demand-driven approach, with the district as the administrative unit with focus on gram panchayats and local communities. It focuses on the infrastructure but equally gives very high priority to behaviour changes. Some key features of the TSC include:

- A community-led approach with focus on collective achievement of total sanitation
- Focus on information, education and communication (IEC) to mobilize and motivate communities towards safe sanitation
- Minimum incentives only for below poverty line (BPL) households/poor/disabled, post-construction and usage
- Flexible menu of technology options
- Development of supply chain to meet the demand stimulated at the community level
- Fiscal incentive in the form of a prize – Nirmal Gram Puraskar (NGP) – to accelerate achievement of total sanitation outcomes

TSC is being implemented at a scale in 607 districts of 30 states/union territories (UTs). Against an objective of 12.57 crore individual household latrines (IHHL), the sanitation facilities for individual households reported to have been achieved is about 8.38 crore as of December 2011. In addition, about 10.32 lakh school toilets, 19,502 sanitary complexes for women and 3.46 lakh Anganwadi (pre-school) toilets have been constructed.

Figure 4.1 shows individual household latrine coverage has more than tripled, from around 22% in 2001 to 67% in September 2012

Though the coverage figures shown above are quite impressive, there are quite a good number of issues that do not get reflected. These are:

(a) The figures above only shows that a number of households have been provided with toilet facilities but do not take into account the sanitary condition of the toilets or its usage.
(b) Improved hygiene behaviour, hand-washing practices with soap, etc. are not getting reflected in these reports. So broad sense of sanitation is not getting reflected.
(c) The Planning Commission's Eleventh Plan Document, page 173, states that not all the toilets built under the programme are being used by all the members of the family for whom the toilets were built.
(d) Studies have shown lower level of toilet usage because of:
    - Inadequate awareness
    - Water scarcity

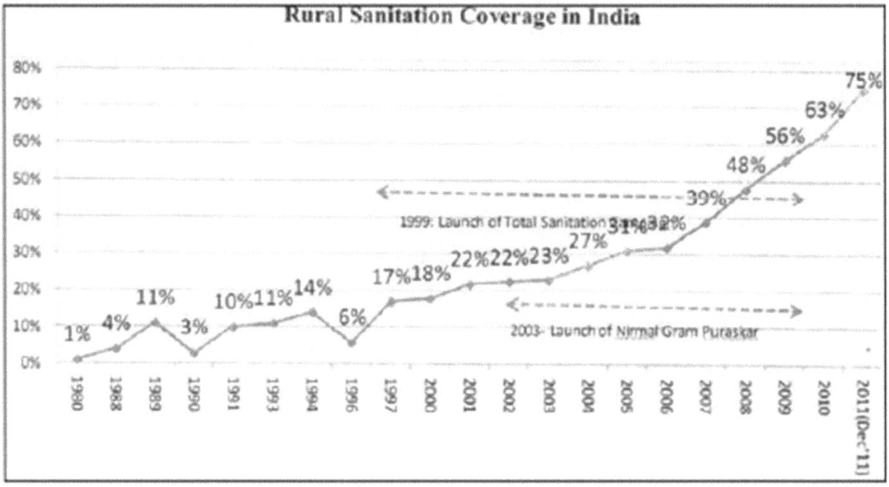

Fig. 4.1 Rural sanitation coverage in India

- Poor construction standards
- Past emphasis on high-cost latrine design

It was always felt that success of the sanitation programme will be achieved only when TSC gets integrated with improved water availability and healthcare programmes. By integrating these programmes, there are possibilities of reducing the incidence of diarrhoea, especially amongst the children and other infectious waterborne diseases. Hence, TSC was slowly integrated with rural water supply programme and National Rural Health Mission (NRHM).

Since TSC also looks for school sanitation and hygiene education programme, it was prudent to converge these programmes with the Department of School Education and Literacy (DSEL) and the Sarva Shiksha Abhiyan (SSA), the flagship programme of the Government of India. The emphasis is to provide the school children with an environment having all facilities and inculcate the improved habit of hygiene behaviour in children.

TSC also introduced the concept of NGP which helped in the introduction of sanitation programme in a big scale in rural India. With this success in view, the TSC was rechristened and named as "Nirmal Bharat Abhiyan" (NBA). The main objective was to accelerate the sanitation programme in rural India with a saturation approach.

The main objectives of the NBA are as follows:

1. Bring about an improvement in the general quality of life in the rural areas
2. Accelerate sanitation coverage in rural areas to achieve the vision of Nirmal Bharat by 2022 with all Gram Panchayats in the country attaining Nirmal status
3. Motivate communities and Panchayati Raj Institutions promoting sustainable sanitation facilities through awareness creation and health education

4. To cover the remaining schools not covered under Sarva Shiksha Abhiyan (SSA) and Anganwadi Centres in the rural areas with proper sanitation facilities and undertake proactive promotion of hygiene education and sanitary habits amongst students
5. Encourage cost-effective and appropriate technologies for ecologically safe and sustainable sanitation
6. Develop community-managed environmental sanitation systems focusing on solid and liquid waste management for overall cleanliness in the rural areas

There are some consequences due to lack of sewerage system or improper functioning of sewerage system. Most of the urban local bodies in the country are either not having any sewerage system with treatment facilities or are not treating the wastewater totally generated by them. So finally, these wastewater find its way into some waterbody like rivers and lakes or spread in the land, causing soil and groundwater pollution. In some cases, this wastewater is being used for agriculture purposes. All of these cause huge problem of both organic and bacteriological pollution with high level of faecal pollution. Studies show that the mean levels of BOD have increased in 6 of the 18 major rivers in the country, accounting for 46% of the total river length nationally.

To achieve sustainable sanitation, more area should be covered under well-maintained piped sewerage system. But there are some constraints in achieving the piped sewerage system. Some of them are discussed here like lack of funds, lack of knowledge about nonconventional sanitation technologies, weak institutions with trained personnel, water shortage and lack of operation and maintenance (O&M).

Water shortage is one of the main constraints in installing and for proper functioning of the sewerage system. A large number of class-I cities in India do not have minimum per capita water supply to sustain the sewerage system. Minimum water supply of 130 litres per capita per day (lcpd) is required to sustain the sewer system in the area. Ahmedabad, Vadodara, Raipur, Rohtak, Hisar, Gurgaon, Bangalore, Mysore, Indore, Navi Mumbai, Imphal, Shillong, Bathinda, Coimbatore, Mathura, Meerut and many more do not have minimum per capita water supply to sustain the sewer system.

After the installation of sewerage system, the proper operation and maintenance is also a big challenge. The existing treatment capacity is also not effectively utilized due to operation and maintenance problem. Some treatment plants are underutilized and some are overloaded. In some treatment plants, the sewage collection is low due to either inadequate sewerage network or the households are reluctant to get them connected to the sewerage system due to high cost of investment. So the treatment units are underloaded. Another problem with the existing sewage treatment plants is majority of them are not conforming to the standards prescribed under the Environmental (Protection) Rules for discharge into streams. Finally, in most of the sewage treatment plants, there is a big gap of availability of trained manpower to run the units. Majority of available personnel know about running the pumping units and hardly know about the sewage treatment processes.

According to census 2011, only 11.9% of total households are covered under piped sewer system. To cover the whole country with sewerage system, a huge investment will be required. As discussed earlier, the sewerage system can only function properly only when there is an adequate piped water supply system with at least 130 litres per capita per day. Below this level, there will not be adequate wastewater to maintain the flow in the sewers. Hence, majority of places will need investment in water projects to meet this need. Apart from this, an investment of Rs.20,000–Rs.60,000 per capita will be needed for the sewerage system, which is beyond the paying capacity of the beneficiaries. Any technology whose total financial investment is more than 10–20% of user income probably should be excluded as financially unaffordable.

Apart from inadequate funds, the developing countries are facing an added disadvantage that their planners and engineers have been trained mostly to handle sewerage systems and not adequately trained to plan, implement and run the nonconventional sanitation technologies. As the priority is to reduce and eventually eliminate the transmission of excreta-related diseases, which can be achieved by having nonconventional sanitation technologies with lower investment, the developing countries should aim to achieve the same. Lack of interest in sanitation technologies other than sewerage is in part because of the standardized education of most planners and engineers in developing countries. People do not know more about the nonconventional sanitation technologies.

All these above-mentioned problems show that to cover the whole country under the sewerage system is not possible in the near future. To achieve sustainable sanitation, it is necessary for our country to go for the nonconventional technologies, as these technologies require less money, less water and less space and do not require skilled labour for operation and maintenance. The technologies which are maintained by the beneficiaries should be promoted because as and when the system collapse, they are able to fix the problem by themselves.

4.2 Issues in Achieving the Sustainable Sanitation Coverage

4.2.1 Access

The public health problems can only be reduced to a great extent by provision of sanitation and clean environment with an adequate quantity of potable water. This can only reduce the incidence of diseases and deaths. Under Millennium Development Goals, the international community wanted to halve the proportion of people without access to safe drinking water and basic sanitation facilities by 2015.

The Joint Monitoring Programme (JMP) for Water Supply and Sanitation published by WHO/UNICEF describes the status and trends with respect to the use of safe drinking water and basic sanitation and progress made towards the MDG drinking water and sanitation target. As we are nearing 2015, it will be interesting to note the countries who have not been able to achieve the coveted target and reasons behind it. The JMP reports (2012) also identify some striking disparities: the gap between progress of providing access to drinking water versus sanitation, the divide between urban and rural populations in terms of the services provided and differences in the way different regions are progressing. However, it should be kept in mind that different countries started from different baselines and there is a vast difference between socio-economic strata in the society.

The census 2011 shows the coverage of sanitation and water supply. The census report shows that 49.8% of 122.9 million households in India practise open defecation. While in the rural India, the situation is still worse. 67.3%, i.e., 113 million households practise open defecation.

4.2.2 Poverty and Disparities

Under the Ministry of Drinking Water and Sanitation (MDWS) programme, an incentive is provided only to below poverty line (BPL) households under the scheme. While the incentive for individual household latrines (IHHLs) has been revised from time to time and stands at Rs. 3200/– (Rs. 3700/– {US$50–60} for hilly and difficult areas) per IHHL constructed and used by BPL household, including state share of Rs. 1400/–, the BPL households are expected to find resources for the remaining cost. Most assessments have calculated IHHL cost at about Rs. 8000/– with the substructure alone costing about Rs. 5000/–. Those who are above poverty line (APL) are expected to be motivated through IEC to construct toilets on their own or through availing of credit facilities.

Apart from these incentives, it has now been decided that sanitation programme activities can be undertaken under MGNREGA in accordance with these guidelines:

(a) Construction of individual household latrines (IHHL) as per instructions/guidelines of "Total Sanitation Campaign" administered by the Ministry of Drinking Water and Sanitation (MDWS)
(b) Construction of Anganwadi toilet unit and school toilet unit as institutional projects
(c) Solid and Liquid Waste Management (SLWM) works in proposed or completed Nirmal Grams

{Cost of unskilled labour (up to 20 man-days) and skilled labour (up to 6 man-days) for material component under Mahatma Gandhi NREGS on construction of individual household latrine. The total amount to be booked under MGNREGA will however not exceed Rs. 4500 per IHHL.}

While the policy of the Government of India under TSC has been to disburse incentives to the BPL households, considered the poorest in the rural areas, poverty continues to be a curse and a barrier for accelerating rural sanitation coverage. This gives an indication of continuing with the practice of incentives to the poor in recognition of their achievement to construct and use sanitation facilities with corrections as may be required to get the intended results.

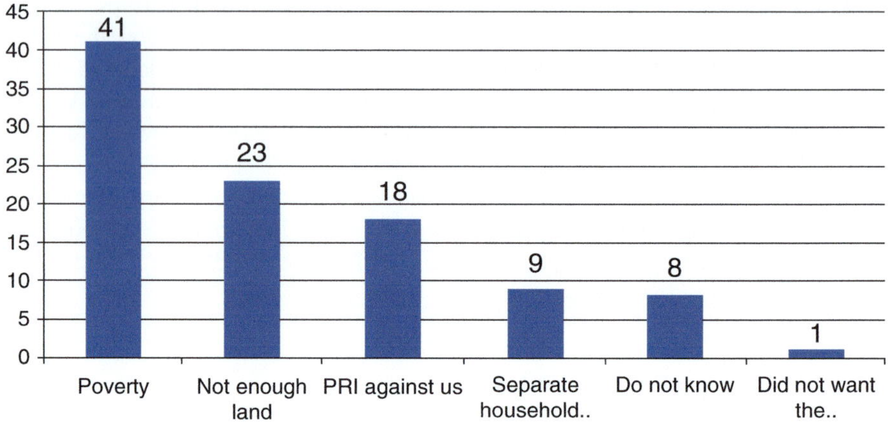

Fig. 4.2 Reason for nonconstruction of toilets

In a study done by Centre for Media Studies (CMS), engaged by the Ministry of Rural Development in the year 2010, 41% of the respondents cited poverty as the reason for nonconstruction of toilets (Fig. 4.2).

4.2.3 Community Approach for Sanitation and Health Benefits

The current allocations are restrictive towards adoption of a community approach to sanitation. An assessment undertaken by the Water Sanitation Programme (WSP) of the World Bank in Himachal Pradesh in 2005 revealed that in villages with approximately 30% sanitation coverage, the incidence of diarrhoea was reported by approximately 38% households. Even in villages with 95% sanitation coverage, the diarrhoeal incidences were reported by around 26% households. Only Open Defecation-Free (ODF) villages with 100% sanitation coverage reported significantly lower incidences of diarrhoea – approximately by 7% households. In effect, even if a few individual households switch to using toilets, the overall risk of bacteriological contamination and incidence of disease continues to be high. To achieve the full goals of sanitation, community saturation approach cutting across the APL/BPL barrier is suggested for creation of Nirmal Grams.

The present approach is to sensitize the community by awareness about the harmful effect of open defecation and lack of sanitation and how it affects the dignity and security of women, children and elderly. The present approach is to provide a cost-effective technology which is ecologically safe and sustainable.

4.2.4 Behaviour Change

Apart from hardware, the biggest challenge is to create and sustain community demand for hygiene and sanitation. Behaviour change does not happen so quickly and may take years to be achieved. But the change is happening to a great extent in the rural areas. Though the country has more or less broken the tradition barrier and taboo associated with toilets, open defecation in rural areas is still being practised both by the rich and poor and is a culturally and socially accepted tradition.

To meet these needs, the information, education and communication (IEC) strategies need to be suitably modified and addressed at different forums to bring about the behaviour change, and this needs to be practised for years till open defecation is totally discarded by the community.

A limitation noted while achieving sanitation coverage is that various field studies have pointed to various levels of latrine usage depending upon the community awareness and also slippage in the status of NGP villages that shows a variable trend. For example, in one such study undertaken by UNICEF in 2008, it was found that out of the 81% of the population having access to sanitation in NGP Panchayats, only 63% were using the facilities.

4.2.5 Septic Tank

Septic tanks are also a big problem in achieving sustainable sanitation target. Majority of the septic tanks had openings into open drains, which drained the liquid effluents from the septic tanks. These open drains either lead to waterbodies and pollute them or cause groundwater pollution wherever these are discharged. Septic tank requires more space. The construction needs regular technical assistance and supervision. This needs ventilation, which adds to cost. De-slugging of septic tanks is needed on a regular basis. The sludge and effluent from a septic tank cannot be used as a fertilizer straight away without causing health hazards. In some areas, septic tank toilets are within a 10-metre distance from water sources.

However, people do demand for septic tanks, as most of the masons available in the rural areas have some knowledge about constructing a septic tank rather than any other safer designs. The masons motivate villagers to go for septic tanks as they had been doing so for generations.

Disadvantages of septic tanks:

- The leaching system is often not constructed and a common practice is to discharge effluent directly into an open drain.
- Septic tanks often receive too much wastewater. As a result, the retention time in the septic tank is insufficient, and the soakaway becomes hydraulically overloaded. Mainly householder bypasses the soakaway and connects the overflow directly to a surface water drain.
- Shock loadings and disturbance of settling zone caused by large inflows (typically from sullage discharges) can affect the efficiency of the septic tank and causes excess solids to flow into the soakaway.
- Performance monitoring of septic tanks is rarely undertaken, and regulation to control private de-slugging operators is problematic. This creates pollution as well as a potential health hazard.

4.2.6 Caste-Based Distribution of Toilets

Construction of household toilets has got significant linkage with caste educational background, economic factor and concept of cleanliness amongst households. Various studies have highlighted these issues. One study conducted by ORG Centre for Social Research shows that more households of the general caste own toilets and there is a significantly lower proportion of SC and OBC households which was found to own household toilets. The survey was conducted in year 2009 under PRWSS World Bank Project in all the districts of Punjab with a sample size of 20 households in each village. The study shows:

- Ninety percent general category households.
- 69.5 percent households in OBC category.
- Fifty-seven percent households in scheduled caste category had this facility.

4.2.7 Dysfunctional Toilets and O&M

The Report of the National Water Supply and Sanitation Committee (1960–1961), Ministry of Health, Government of India, had recommended that the district centre would be the pivot of future activity for implementing the rural programme in the future. It is desirable that the centre is so developed that it has a manufacturing year for casting, curing and storage of different sizes of concrete pipes, specials, latrine pans, squatting slabs, traps, etc. required for rural water supply and sanitation works. A mechanical section under adequate supervision of the district centre should be entrusted with the production, stocking, supply and distribution of all materials required for the programme. Each centre could, in addition, train the required number of masons, carpenters, mechanics, mysteries and other artisans in their respective works so that they may handle the field work in rural areas to better advantage. The district centre could, in addition, arrange for the necessary orientation, refresher and training courses for the subordinate technical personnel employed in the rural areas to have a better advantage.

Though important suggestions for implementation of O&M in rural sanitation were made as early as 1960–1961, not much action have been initiated in the last five decades in taking up these important issues of the programme seriously in the districts.

In the initial stages of the programme, the funding for the construction of household toilets was quite low, and there was lack of proper monitoring of the toilet construction process. This resulted in large-scale failure of the toilet constructed during this period. To attain Nirmal Gram objectives, it will be necessary for a policy intervention to ensure that these may be made functional and appropriate. There is also a need to evolve a proper O&M mechanism for institutional and community toilets (Fig. 4.3).

Various studies conducted from time to time show the following gaps between the toilet constructions, programme and its uses by the beneficiaries. The main reasons for this gap, which was as high as 51% of toilet constructed and functional, are:

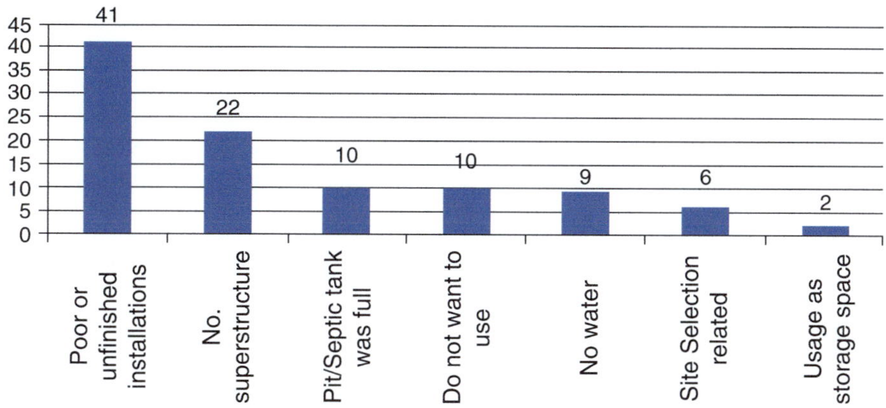

Fig. 4.3 Major reason for dysfunctional toilets

(a) Poor quality of construction material
(b) Dysfunctional toilets for reasons like pit/septic tank full and chocked pan/pipes
(c) Poor site location
(d) Toilet room filled with debris
(e) Toilet room used as storage space

4.2.8 Institutional Framework with Participation of NGOs

TSC Guidelines recommends for a big role for the non-governmental organizations (NGOs) for planning, implementation, motivation and advocacy for rural sanitation programme in the country. Since behaviour change is a very important factor for the success of the programme, the NGOs with their grass-root level functionaries can play a vital role.

It is now recognized that programmes impacting social practices require greater involvement of civil society and its organizations. Local self-help groups, women's organizations, youth associations and NGOs of repute can play a major role in programme implementation. NGOs can contribute immensely in ensuring sustainability of ODF status and monitoring apart from demand generation, resource mobilization and capacity building of stakeholders. Appropriate mechanisms need to be built for them to be encouraged to engage in the sanitation sector.

4.3 Conclusion

To achieve sustainable sanitation coverage in rural India is a big challenge faced by the State Governments and Government of India. Despite initiatives taken by the Government of India, the sanitation coverage is very low as per census 2011. There are many causes behind the low sanitation coverage, but some of them are discussed in many studies like access to the facilities, poverty and disparities, behaviour towards sanitation, caste-based distribution of toilets, dysfunctional toilets and O&M, etc. One of the main reasons is dysfunctional toilets constructed by earlier programmes funded by the government agencies, and the beneficiaries find it difficult to rebuild them and hence are not in a position to use them.

A.K. Sengupta is an environmental engineer with more than 45 years of professional experience in the field of water supply, environmental sanitation, solid and liquid waste management, health and hygiene issues. Working as director general with Sulabh International Academy of Environmental Sanitation and Public Health, he guided project planning, implementation, operation and monitoring; training programmes concerning drinking water, sanitation and solid and liquid waste management; and development of reports. Working with WHO for around 9 years, as national professional officer and sector coordinator for Sustainable Development and Healthy Environments Cluster in India office, he supported planning and implementation of studies/pilot projects concerning the water supply and sanitation sector, water quality issues, wastewater management, strengthening sanitation practices, water safety plan, indoor and outdoor air pollution, hazardous chemicals, food safety, climate change and its impact on health, healthy city programme and occupational health. Apart from WHO, he worked as a consultant with UNICEF, UNIDO and the World Bank for 6 years and provided inputs for the development of the water and sanitation sector in the country and South Africa. Working as deputy adviser/assistant adviser with the Ministry of Urban Development/Ministry of Rural Development, GOI, for more than 15 years, he provided technical assistance and guidance for the planning and implementation of urban/rural water supply projects in the eastern, southern and northern states of India. He organized training programmes on various water and sanitation-related subjects for staff members of state departments, NGOs and private agencies.

- He led development of the concept paper of HRD for the rural water and sanitation sector and assisted in the development of the International Training Network in India.
- He was involved in the development and installation of community-based water treatment plants and water quality surveillance mechanism in the rural areas.
- He was involved in the development of low-cost sanitation programme in rural India. This included the development of appropriate technology, its application and community-based implementation, operation and monitoring of the system.
- He was involved in framing the country's seventh and eighth Five-Year Plan documents for the water and sanitation sector.
- He assisted the state agencies with their emergency water supply and sanitation programme during drought and floods and participated as member of the Central Relief Assessment team.

Apart from GOI, he worked with the Public Health Engineering Department of the Government of Rajasthan for more than 10 years for planning, implementation, operation and maintenance of both urban and rural water supply schemes.

Development of Guideline for Implementation of Water Safety Plan for the Rural Water Supply Systems in India

5

K.J. Nath and A.K. Sengupta

5.1 Introduction

5.1.1 Rural Water Supply Scenario in India: Safety and Quality Issues

Water borne disease remains one of the major health concerns in the world. Diarrhoeal diseases, which are largely caused by contaminated drinking water sources and inadequate sanitation facilities, account for 2.4 million deaths each year and contribute over 73 million disability-adjusted life years worldwide. India with 800 million people in 1.6 million rural habitations spread out 15 diverse ecological regions is no exception. As estimated a couple of years back, around 25 million Indians were affected by waterborne diseases annually with around 700,000 (mostly children below 5 years) death due to water and inadequate sanitation.

In the period prior to independence, the rural community were mostly dependent on open wells and ponds to meet their drinking water needs. People were aware of the pollution and hence made efforts to isolate these sources as far as possible. However, with the passage of time and increase in water demand due to population rise, people had to go for any water source they found to meet their drinking water needs. As most of these sources were grossly contaminated with pathogenic organisms, the burden of communicable diseases like diarrhoea, cholera, typhoid, hepatitis, worm infections, etc. was rampant, and epidemics occurred frequently. Infant

K.J. Nath (✉)
Institution of Public Health Engineers (India),
IPHE Bhawan, CK-58, Sector-II, Salt Lake City, Kolkatta 700091, West Bengal, India
e-mail: nathkj@hotmail.com

A.K. Sengupta
Alumni Affairs and External Relations, Indian Institute of Engineering Science and Technology, Howrah, Shibpur 711103, India
e-mail: ak.sengupta48@gmail.com

K.J. Nath, V.P. Sharma (eds.), *Water and Sanitation in the New Millennium*,
DOI 10.1007/978-81-322-3745-7_5

mortality rate was very high. As disease burden due to diarrhoea deaths and manpower loss was disproportionately high, the Government of India initiated exploration of groundwater by a strong campaign for providing handpump-operated tube wells, thereby providing bacteriological safe water to the rural community. Presently, around 85% of rural population is dependent on groundwater as drinking water source. Though bacteriological contamination of drinking water got reduced substantially, with more and more exploration of groundwater, the chemical contamination of sources crept in more or less unnoticed initially, but now with better water quality testing mechanism a broad mapping has been done. The chemical contaminates mainly include arsenic, fluoride, iron, nitrate, salinity, etc. These have added a new dimension to the problem. To solve this problem, more and more piped water supply schemes with proper water treatment are getting introduced for the rural community in the country.

5.2 Implementing Water Safety Plan in Rural Water Supply Systems

The traditional approach to water quality and safety management in the rural water supply system in the country presently is:

- Water quality test as the source gets developed initially
- Water quality test as it leaves the water treatment plant done occasionally
- Water quality test in the water supply system by collecting water from selected points This again is done occasionally to test the quality of water available to the community.

This is what is referred to as 'end-product testing'. The problem with this approach is that the results are too little and too late for preventive action as water has already been consumed by the community and damage has already been done. With these issues in view, it will be necessary to introduce water safety plan (WSP) in the rural water supply systems in the country to make water safe and potable.

WSP is an improved risk management tool. It identifies:

- The hazards that the water supply is exposed to and the level of risk associated with each
- How each hazard will be controlled
- How the means of control will be monitored
- What actions are required to restore control
- How the effectiveness of the whole system can be verified
- With diverse demographic, socio-economic character, ecological and hydrological conditions prevailing in the country along with different institutional and infrastructure capacities of rural water supply service sector, planning and implementation of a comprehensive WSP for the rural India is an enormous challenge.

The revised National Rural Drinking Water Programme (NRDWP) Guidelines 2009–2012 issued by Rajiv Gandhi National Drinking Water Mission, Ministry of Drinking Water and Sanitation, Government of India, has shifted the focus from 'source development and installation of water supply system for providing drinking water supply to rural household' to focus on development of 'village security plan' which also includes village safety plan before taking up planning and installation of water supply system to ensure provision of safe and adequate water supply to each rural household at a convenient location on a sustainable basis. Basically it envisages provision of drinking water as a part of the overall water resource management and safety plan. In the above context Sulabh International Academy of Environmental Sanitation and Public Health, supported by WHO India Country office, undertook an exercise to review the safety issues and identify and assess the hazards and risks, associated with different rural water supply systems both spot sources and piped water supplies and developed guidelines for implementing water safety plan for the same.

This chapter discusses the implementation of water safety plan for the rural water supply systems which are being used in varying extents by the rural communities in different regions of the country. Though it will be difficult to discuss all the systems, the systems at Sl. No. I to XII have been considered for this chapter.

1. Dug well-based rural water supply system
2. Pond-based rural water supply system with appropriate treatment and rainwater harvesting system through surface storage
3. Bore well-based rural water supply system
4. Roof top rainwater harvesting system
5. Groundwater recharge system
6. Gravity-fed water supply systems for rural communities particularly in the hilly areas
7. Arsenic removal systems for groundwater-based schemes
8. Fluoride removal systems for groundwater-based schemes
9. Iron removal systems for rural schemes
10. Disinfection systems for rural schemes
11. Pump and tank systems for single village
12. Piped water supply systems with appropriate treatment for multiple villages

5.2.1 Critical Constraints and Key Challenges

Given the demographic and socio-economic character, ecological and hydrogeological diversities of rural India and the institutional and infrastructural capacities of rural water supply service sector, planning and implementation of a comprehensive water safety plan for the rural water supply systems is a big challenge. Key constraints and concerns include the following:

(i) *Limited data availability*

Basic data in respect of catchment and source water quality and also various components of water transmission, distribution, storage and household handling in the rural areas is difficult to be obtained. Assessment of risks from chemical and microbial contamination of the system from catchment to consumer end becomes difficult in the absence of regular sanitary inspection data and information.

(ii) *Unplanned development*

Limitations in regulatory institutions and enforcement mechanism have resulted in unplanned development and fragmented responsibility which make it difficult to locate water supply lines and hazards to the same.

(iii) *Sanitation*

Poor status of rural sanitation including facilities for human excreta disposal drainage and solid waste management and also open defecation in many areas increases the risk of faecal contamination of the system in the rural areas.

(iv) *System knowledge*

Because of lack of culture and expertise in community water supply management in most of the village level organizations in the rural areas, much of the information of the piped network may not be available due to poor record keeping or lack of post construction documentation.

(v) *Equipment/human resource availability*

In the time of setting up of health-based standards for appropriate water quality parameters, both microbial and chemical, one should carefully consider the availability of resources as well as infrastructure such as laboratories, skilled manpower, technical expertise, etc. The planned WSP for achieving the same must also address these problems, including capacity building and resource mobilization.

5.2.2 Institutional Mechanism at the State, District and Block/Village Level

Provision of water and sanitation is a state subject. Thus, all programmes related to drinking water supply are implemented by the states. As per 73rd amendment of the Indian constitution, most of the rural development programmes including provision of water supply and sanitation in rural areas is now implemented by the three-tier Panchayet system (rural local government), under National Rural Water Supply Programme (NRWSP). All states are also required to provide drinking water facilities in all the rural schools and ICDS centres from fund available under NRWSP.

To plan and implement a comprehensive WSP in all the rural water supply schemes, it is now required that (i) all the existing rural institutions are firmly committed to the concept of WSP, health-based targets, risks and hazard analysis, operational control, preventive measures, sanitary inspection, quality surveillance, etc.; (ii) additional financial resource is granted to the executing agencies; (iii) capacity building and training in terms of additional manpower, laboratories, testing kits, etc.

are undertaken; and (iv) an effective coordination mechanism is developed between PRI organizations and Health and PHE Department, at the state, district, block and village level for implementing WSP.

5.2.3 Formation of WSP Teams

Figure 5.1 depicts the responsibilities of various level institutions in the context of planning and implementation of WSP. Instead of the creation of new institutions for implementation of WSP, strengthening of the existing institutions with additional manpower, laboratory infrastructure, etc. would be desirable. However, as indicated in Fig. 5.1 at the scheme level, it might be required to have a support team for overseeing the implementation of WSP.

The Public Health Engineering Department (PHED) is the nodal Department for implementing major rural water supply systems involving large water treatment plants and multiple village transmission cum distribution systems. PHED is also responsible for water quality monitoring and surveillance at the state, district and block level. Panchayati Raj organizations (PRI) are responsible for operation and maintenance of spot sources and also village level distribution systems.

5.2.4 Case Studies on WSPs for Rural Water Supply Systems in India

Of the various water supply systems in practice in the rural areas of the country, WSP for two typical systems have been depicted in the following sections.

5.2.4.1 Example I: Water Safety Plan for Dug Well-Based Rural Water Supply System

Source: The rainwater percolates through the soil pores and is deposited as groundwater. This water is generally free from suspended particles. However, chances of chemical contamination are there if ground strata through which the rainwater is percolating contain soluble chemicals. Chances of bacteriological contamination are also very high. This water is collected traditionally by constructing dug well.

Treatment: Dug wells can often supply drinking water at a very low cost, but because impurities from the surface easily reach dug wells, a greater risk of contamination occurs for these wells when they are compared to deeper tube wells. Dug wells are relatively easy to contaminate, and dug wells are unreliable. Basically water of dug well is not treated except disinfection. Disinfection is accomplished by adding chlorine depending on the demand. Usually bleaching powder is used as the source of chlorine. Pot chlorination is generally adopted for dug wells.

Transportation: On lifting of water from the dug well, it is collected in earthen or metal pitchers which are carried by the villagers to their respective homes and finally stored in the kitchen/dining in a bigger pitcher (Jala) for consumption.

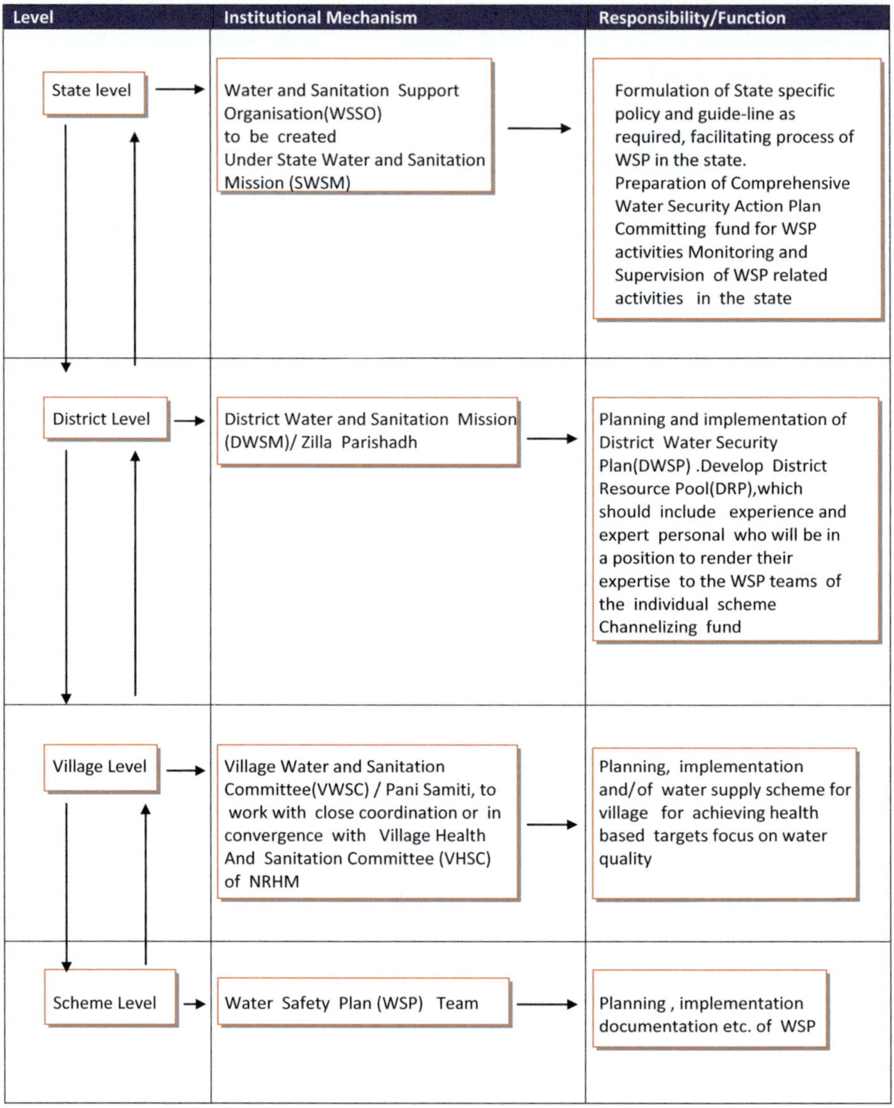

Fig. 5.1 Institutional mechanism at the State/District/Village level

Hazard Event

In case of dug well the following hazard events have been identified as which need proper attention to ensure safety of the water:

(i) Direct entry of contaminated water in the well
(ii) Contamination due to damage to lining of dug well
(iii) Accumulation of contaminated water due to poor drainage arrangements
(iv) Contamination from the buckets
(v) Contamination due to naturally occurring chemicals
(vi) Leaching of chemicals
(vii) Leaching of microbial contamination

Broad control measures to be adopted for the above mentioned events are:

(i) Proper construction of well including parapet wall, cover, apron and drainage.
(ii) Proper sealing of lining with good mortar.
(iii) Extension of drains to a nearby nullah or ditch.
(iv) Installation of handpump.
(v) Discard the source and select one without chemical contamination.
(vi) Hazardous chemicals shall not be dumped near the dug well.
(vii) Increased length of travel to prevent contamination.

For each of the above hazard event, it causes hazard type or risk involved, and control measures required to be adopted for prevention of the hazards in respect of dug well are indicated in Tables 5.1, 5.2 and 5.3.

5.2.4.2 Example II: Pump and Tank Water Supply for Single Village

There are many ways to cover rural population under safe drinking water supply. Though spot sources are the common method to provide water in rural areas, piped water supply schemes are also being implemented. For uncovered habitations, one of the options in those areas could be pump and tank water supply systems. Such schemes are small and can be habitation specific. Local community can also easily implement and maintain such schemes without much external support. In this system, a small drinking water source is to be created first, and then water is to be supplied to the villagers by pumping. Considering need of the villagers, either single tank near the water source or multiple tanks at different locations of the habitation could be provided in such water supply systems. In case of providing multiple tanks, distribution lines are to be laid to cover the entire habitation.

Tables 5.4, 5.5 and 5.6 depicts the hazard analysis, operational monitoring, surveillance and verification.

Table 5.1 Hazard analysis (dug well)

Process step	Hazardous event	Hazard type	Risk prioritization (Semi-quantitative method)			Control measures	Additional control measures
			In category of likelihood	In category of consequence/impact	Risk rating		
Well construction	Ingress of contaminated surface water directly into the well as lining stops at ground level	Construction	Moderate (3)	Major (4)	3 × 4 = 12	Extend lining up to a height of 0.9 m	
	Ingress of contaminants due to poor construction or damage to lining	Construction	Moderate (3)	Minor (2)	3 × 2 = 6	Repair the leaky and damaged joints	
Catchments around dug well	Leaching of organic matter from organic waste damps	Microbial	Unlikely (2)	Catastrophic (5)	2 × 5 = 10	Locate waste dumps at least 10 m away from dug well	
	Leaching of faecal matter from human and animal wastes	Microbial	Moderate (3)	Major (4)	3 × 4 = 12	Locate toilet and animal pen at least 10 m away from dug well	Chlorinate the dug well periodically with bleaching powder
	Leaching of naturally occurring chemicals in subsoil strata	Chemical	Minor (2)	Major (4)	2 × 4 = 8	Discard a well having chemical contamination beyond permissible limit	

Dug well	Contamination from the use of insanitary bucket and rope	Microbial	Moderate (3)	Major (4)	$3 \times 4 = 12$	Ensure use of dedicated set of bucket and rope for lifting water. The bucket shall always be kept on a raised platform	The bucket shall be cleaned and disinfected daily
	Contamination due to inundation during flood	Microbial	Minor (2)	Major (4)	$2 \times 4 = 8$	The platform shall be adequately raised and well lining shall be above the flood level and adequate drainage to prevent water logging	Post flood super-chlorinate
	Contamination due to use of poor quality water for priming of handpumps attached to a dug well	Microbial/chemical	Moderate (3)	Major (4)	$3 \times 4 = 12$	Leaky sheet valve to be replaced immediately on detection. In emergency use only good quality water for priming	
Transportation of water	Contamination due to dust if water container is not properly covered during transportation	Microbial/others	Moderate (3)	Major (4)	$3 \times 4 = 12$	Ensure covering the container	The carrying container shall be cleaned before use

(continued)

Table 5.1 (continued)

Process step	Hazardous event	Hazard type	Risk prioritization (Semi-quantitative method)			Control measures	Additional control measures
			In category of likelihood	In category of consequence/impact	Risk rating		
Storage and handling of water	Contamination due to the use of unclean utensil and/or hand for drawing water from storage tank (Jala)	Microbial/others	Moderate (3)	Major (4)	$3 \times 4 = 12$	Water shall be taken out from the storage tank using a cup fitted with a handle	Make it a habit to wash hands before touching the water pitcher or storage tank
	Contamination by dust particles or domestic animals for not covering the storage tank	Microbial/others	Moderate (3)	Major (4)	$3 \times 4 = 12$	Storage tank shall always be kept covered	Regular washing of storage tank to be practiced

Table 5.2 Operational monitoring schedule

Process step	Performance indicators	Monitoring process		Critical limit	Corrective action		Supporting programme
Catchments around dug well	Waste dumping, toilet, etc. close to dug well	What	Presence of waste dump, latrine, etc. around dug well	No latrine, animal pen, waste dump, etc. within 10 m radius of dug well	What	Resitting of the units	IEC activities
		How	Sanitary inspection		How	Applying GP rules	
		Where	Within 10 m radius of dug well		When	Immediately on identification	
		When	Once in every 6 months		Who	Member of GP	
		Who	Chemist of NGO Lab				
Dug well	Quality of water of the dug well	What	Test quality of water	Water quality must satisfy the standard set forth in water supply manual of GOI	What	Unsafe wells shall be marked	Educate villagers through group meetings and IEC activities
		How	Laboratory test		How	Through routine testing as per GOI guidelines	
		Where	At source		When	Immediately on detection	
		When	During commissioning and periodically at least once in a year		Who	SAE PHED/Chemist of NGO Laboratory	
		Who	Concerned laboratory				
	Condition of well lining and drainage around dug well	What	Prevention of leakage through lining and inundation	Lining and drains are in good condition	What	Sealing of damaged lining and repairs of drains/platform	
		How	Sanitary inspection		How	Using cement mortar	
		Where	Dug well lining and platform/drain		When	Immediately on detection	
		When	Once in every 6 months		Who	SAE PHED placed at block level	
		Who	Chemist of NGO Lab				
	Disinfection of dug well	What	Disinfection using bleaching soln.	Smell of chlorine	What	Chlorine dose as fixed by lab	Training by laboratory on disinfection
		How	Using field test kit		How	By pot chlorination	
		Where	At dug well		When	Twice in a week	
		When	Daily		Who	Caretaker	
		Who	Caretaker				

(continued)

Table 5.2 (continued)

Process step	Performance indicators	Monitoring process		Critical limit	Corrective action		Supporting programme
Handpump	Maintenance of pump	What	Checking operation of pump	Pump seat valve is leaking	What	Repairing of pump	
		How	By operating the pump		How	By replacement of seat valve	
		Where	At dug well site		When	Immediately on detection	
		When	Monthly		Who	Mechanic PHED	
		Who	Mechanic PHED				
Storage and handling	Contamination during handling and storage	What	Hygiene practice during collection, transport and storage	Water collection, transportation and storage are hygienic	What	Key hygiene messages	Educate villagers through group meetings and IEC activities and school children as a part of curricula
		How	Through inspection at all levels		How	Hygiene education and practice at schools and IEC activities	
		Where	At dug well site and households		When	Regular activity	
		When	Weekly		Who	Member of GP	
		Who	Member of GP				

Table 5.3 Surveillances and verification schedule

Activity	Description	Frequency	Responsible party	Records
Sanitary survey	Sanitary survey will be done as per GOI guidelines	Twice in a year	Chemist of NGO Lab	NGO Laboratory and GP office
Water quality monitoring	Monitoring of quality of water for physical, chemical and biological quality as per GOI guidelines	Twice in a year	Chemist of NGO lab	NGO Laboratory and GP office
Effectiveness of water safety plan	Monthly meeting of the Water and Sanitation Committee shall discuss the WSP involving the District Water and Sanitation Cell, PHED and BMOH and meet the community	Field visit and discussion with the communities to be done as frequently as possible. Gap shall not exceed 3 months	Gram Panchayat	GP office
Impact assessment of advocacy programme	Verification of water-related practices of the community	Random sampling	Chemist of NGO Lab/GP office Staff	IEC material produced (NGO Laboratory and GP office)
Checking satisfaction level of the user community	Visit to households and meeting at GP office	Quarterly	Panchayat/GP office	Survey report (GP office)

5.2.5 Conclusions

Safe water and sanitary disposal of human excreta have been central to the concept of public health. To protect and promote community health, public water supply has to be in sufficient quantity (for maintaining a decent standard of personal and home hygiene), of adequate quality (safe for human consumption) and accessible to all segments of the community. In India, like many other countries, waterborne diseases still occur in a big way. Statistics do show that more than 7.5% of all environment-related deaths in the country are due to water sanitation and hygiene-related diseases. Annual DALYs lost is around 9.4%. Outbreaks show that one cannot solely rely on water treatment indicator because end point testing is too little and too late. Hence, the concept of water safety plan as advocated by WHO needs to be adapted for providing safe water supply to the community.

Table 5.4 Hazard analysis*

Process step	Hazardous events	Hazard type	Control measures	Risks	Additional control measures
Construction of tube well	Presence of arsenic, fluoride, iron, salinity or any other toxic materials beyond permissible limit	Chemical contamination of water	Testing of water and discarding water source if necessary	Toxic effect on human	Use of toxic contaminant removal plant
Construction of tube well	Poor yield from tube well	Non-availability of water	Selection and tapping of appropriate aquifer	Insufficient yield of water affecting supply	Geohydrological investigation yield test of tube well
Construction of tube well	Contamination of tube well water at ground level	Bacteriological contamination	Provision of proper platform, drainage and soakage pit	Suffering of people from diseases	Providing cover and sanitary sealing of tube well Adequate distance of tube well from latrines, atleast around 10 meters distance to be maintained, etc.
Repair of pump	Contamination of tube well water	Bacteriological contamination	Disinfection of pump and spares before lowering	Supply of bacteriological contaminated water	Disinfection of tube well water
Supply of water	Non-availability of electricity	Disturbance of supply	Provision of stand by generator	Non-availability of water by the villagers	Regular contact with electricity department
Supply of water	Non-functioning of pump	Damage of pump	Keeping ready stock of spares	Non-availability of water by the villagers	Availability of trained local mechanic
Supply of water	Disturbance in distribution system	Leakage or any other problem in pipe lines	Keeping ready stock of spares	Non-availability of water by the villagers	Availability of trained local plumber
Use of raw water for treatment	Contamination of surface water body	Contamination of supplied water	Stop rough use of the water by the villagers	Supply of contaminated water	Construction of embankment to arrest run off IEC campaign

Treatment of surface water	Improper functioning of treatment plant	Insufficient and poor quality of treated water	Routine check and maintenance	Non-availability of water by the villagers	Maintenance of stock of spares
Supply of water	Leakage in distribution system	Bacteriological contamination	Keeping ready stock of spares	Supply of contaminated water	Availability of trained local plumber Disinfection of distribution system
Supply of water	Stoppage of water supply	Disturbance in supply	Availability of trained pump operator	Non-availability of water by the villagers	Availability of local standby operator
Maintenance of scheme	Availability of fund	Disturbance in supply	Collection of water tariff from villagers	Non-availability of water by the villagers	Availability of fund from Panchayat

*In the above table hazardous event and risks have been described subjectively. However, it is preferable that in line with Table 5.4 of Chap. 3, the WSP team could quantify and prioritize risks

Table 5.5 Operational monitoring schedule

Process step	Performance indicators	Monitoring process	Critical limit	Corrective action	Supporting programme
Quality of drinking water	Parameters of drinking water	Testing of chemical quality and regular testing of bacteriological qualities	Within permissible limit as per BIS10500	Creation of alternate source in case of chemical contamination	Ensure availability of services from the water testing laboratory
				Disinfection in case of bacteriological contamination	If necessary, capacity of laboratories to be augmented
Pumping of water	Operation of pumps (hours)	Maintenance of log book for pump operator and pump	Operation time of pump to be ensured to produce and supply designed quantity of water (hours)	Vigilance over duty of the pump operator and action to ensure that the pump is always in running condition	Engagement of local people as pump operator and training of pump operators.
					Make spare parts available
Sanitary measures to protect water source	No bacteriological contamination	Regular testing of chemical and bacteriological quality of supplied water	Within permissible limit as per BIS10500	Disinfection in case of bacteriological contamination	Provide platform, drains, soak pit, etc. at tube well site
	Presence of free residual chlorine				Protective measures surrounding surface water body. Device for measuring residual chlorine
					Involve community IEC campaign
Process step	Performance indicators	Monitoring process	Critical limit	Corrective action	Supporting programme
Collection of water by villagers	Availability of required quantity of water for every family on time	Feedback from villagers	40 lpcd for each family	Ensure duration of supply	Involvement of users and launching of IEC activities
				Proper maintenance of distribution system	
Stoppage of wastages of water	No leakage in pipe lines and water tanks	Involvement of users	Availability of water for each of the family	Timely repair of pipe lines, taps, etc.	Involvement of users and launching of IEC activities
	Taps remain closed when not used				

Table 5.6 Surveillance and verification schedules

Activities	Agency responsible for verification	Frequency of verification	Method of verification and how records would be maintained
Functioning of water supply scheme	VWSC/VHC	Once in a month	With the help log book maintained for pumps Feedback from the users based on format Records are to be maintained by VWSC
Availability of desired quantity of water	VWSC/VHC	Once in a month	Feedback from the villagers based on a prescribed format
Quality of water	VWSC/VHC and facilitator (water sample collector)	Once in a month	With the help of nearest water testing laboratory and residual chlorine measuring kit Records are to be maintained by the VWSC
Availability of fast- and slow-moving spares	GP, pump operator, plumber and mechanic	Once in a month	Verification format to be developed by the SAE of the block
Collection of water tariff, opening of bank account and maintenance of cash book	GP Pump operator to collect water tariff Two authorized persons to sign cheques	Checking by VWSC Collection in every month Audit once in a year	To be verified by the representative of concerned GP To be audited every year by external auditors

Acknowledgements The chapter is based on the guidelines for rural water supply systems developed by Sulabh International Academy of Environmental Sanitation and Public Health (SIAES), supported by WHO, India, and with the help of an expert group consisting of reputed experts from all over the country. The authors acknowledge the services provided by the project team and management of SIAES with sincere thanks.

References

1. http://www.whoindia.org/LinkFiles/Water_Quality_Guidelines_for_Water_Safety_Plans_for_Rural_Water_Supply_Systems_pdf
2. http://www.wsportal.org/uploads/IWA%20Toolboxes/WSP/WS%20conference/pacific%20hall/KJNath%20WS2012.pdf
3. http://wsp.org/sites/wsp.org/files/publications/Water-Safety-Plans-rural-water-supply-India.pdf – Water Safety Plan for Rural Water Supply in India, Policy Issues Institutional Arrangements – Report by WSP-World Bank, November 2010
4. http://www.who.int/water_sanitation_health/dwq/wsp170805chap13.pdf – Report by WHO
5. http://www.indiawaterportal.org/sites/indiawaterportal.org/files/Water%20Safety%20Plans_Rural%20Water%20Supply%20Systems_SIAES_WHO_2010.pdf
6. http://ipheindia.org/yahoo_site_admin/assets/docs/scan_1.112164717.pdf

Prof. K.J. Nath, BE (Cal), MEPH (Cal), DSE (Delft). FIE, FNASc, FIPHE, FIWWA, FNESA is widely regarded as one of India's foremost experts on sanitation and community water supply, environmental health and public health engineering. As professor of environmental sanitation and director of All India Institute of Hygiene and Public Health (AIIH&PH), Government of India (GOI), he did seminal works on rural and urban sanitation and water supply, urban waste management, air and water pollution, water quality surveillance, environmental epidemiology and impact assessment. His expertise has been requisitioned by several national and international organizations including the Planning Commission, Council of Scientific and Industrial Research (CSIR), Indian Council of Medical Research (ICMR) Institutes and Ministries of Environment, Health, Urban Development and Rural Development of the Government of India as well as state governments, local and development agencies and research academic institutes. At the international level, he has served on a number of occasions as a consultant and member/chairman of expert panels of the World Bank; Asian Development Bank; WHO; UNICEF; UNEP; UNDP; WSSCC; UN-Habitat; International Scientific Forum on Home Hygiene (IFH), UK; U.S. Trade and Development Agency; etc. He served as a member of the Steering Committee/Working Group of the Planning Commission from the sixth Five-Year Plan to the tenth Five-Year Plan. He was a member of the Planning Commission High Power Committees on SWM and Public Health (1997–1998) and Steering Committees of the 1st- and 2nd-Phase Ganga Action Plan, an expert member of NGRBA (2009–2014) and the National Task Force on Environmental Health and the chairman of the National Co-ordination Council on Water Quality and Rajiv Gandhi Drinking Water Mission. Presently, he is the **chairman of the Arsenic Task Force, West Bengal; chairman of the Core Committee on Water Quality, Safety and Security, Government of West Bengal; member of the Scientific Advisory Board; SEA Regional Coordinator of the International Scientific Forum on Home Hygiene (IFH), UK; member of the Technology Advisory Group, WHO (WASH Climate Resilient); and president of the Institution of Public Health Engineers, India (IPHE),** the premier national professional association of the public health engineers in India. He has published more than 150 research and review papers, books, monograms, etc. He has received several awards including the Rajiv Gandhi Memorial Sulabh Sanitation Award (1998), Environmental Scientist of the Country award by National Environmental Science Academy (1999), Golden Jubilee Award of the Indian Public Health Association (2007) and Most Eminent Public Health Engineer of the Country by the Institute of Engineers in 2008. **He is a fellow of the National Academy of Sciences, India; Indian Public Health Association; Institution of Public Health Engineers, India; Institution of Engineers; and Indian Water Works Association, and he is an honorary fellow of the National Environmental Science Academy, India, and member of the International Water Academy, Oslo. Prof. Nath has to his credit, more than 150 publications of books, monograms, course manuals, etc. and papers presented in International and National Conferences and published in reputed International and National Journals.**

A.K. Sengupta is an environmental engineer with more than 45 years of professional experience in the field of water supply, environmental sanitation, solid and liquid waste management, health and hygiene issues. Working as director general with Sulabh International Academy of Environmental Sanitation and Public Health, he guided project planning, implementation, operation and monitoring; training programmes concerning drinking water, sanitation and solid and liquid waste management; and development of reports.Working with WHO for around 9 years, as national professional officer and sector coordinator for Sustainable Development and Healthy Environments Cluster in

India office, he supported planning and implementation of studies/pilot projects concerning the water supply and sanitation sector, water quality issues, wastewater management, strengthening sanitation practices, water safety plan, indoor and outdoor air pollution, hazardous chemicals, food safety, climate change and its impact on health, healthy city programme and occupational health. Apart from WHO, he worked as a consultant with UNICEF, UNIDO and the World Bank for 6 years and provided inputs for the development of the water and sanitation sector in the country and South Africa.Working as deputy adviser/assistant adviser with the Ministry of Urban Development/Ministry of Rural Development, GOI, for more than 15 years, he provided technical assistance and guidance for the planning and implementation of urban/rural water supply projects in the eastern, southern and northern states of India. He organized training programmes on various water and sanitation-related subjects for staff members of state departments, NGOs and private agencies.

- He led development of concept paper of HRD for the rural water and sanitation sector and assisted in the development of the International Training Network in India.
- He was involved in the development and installation of community-based water treatment plants and water quality surveillance mechanism in the rural areas.
- He was involved in the development of low-cost sanitation programme in rural India. This included the development of appropriate technology, its application and community-based implementation, operation and monitoring of the system.
- He was involved in framing the country's seventh and eighth Five-Year Plan documents for the water and sanitation sector.
- He assisted the state agencies with their emergency water supply and sanitation programme during drought and floods and participated as member of the Central Relief Assessment team.

Apart from GOI, he worked with the Public Health Engineering Department of the Government of Rajasthan for more than 10 years for planning, implementation, operation and maintenance of both urban and rural water supply schemes.

Household Water Treatment: Health Significance and Risk-Based Approaches for Consumer Safety

6

Nirmala Ronnie, Peter McClure, and Nimish Shah

6.1 Introduction

Diarrheal diseases resulting from consumption of contaminated drinking water are recognized as a major contributor to mortality and morbidity in developing countries. In their recent paper on global access to safe water, Onda et al. [1] present revised estimates of populations accessing safe water and predict that up to 3 billion people may either lack access to safe water or may be accessing water that could be a sanitary risk. Despite good progress in providing access to improved sources of water, the study estimates that as many as 1.2 billion people who have access to improved sources may still be using unsafe water with significant sanitary risks (Fig. 6.1).

Increased coverage of sanitation and improved sources of drinking water has contributed to reduced disease burden and the associated mortality and morbidity. The Joint Monitoring Programme (JMP) on Millennium Development Goals (MDGs) classifies water sources as unimproved and improved. Unimproved sources have been segregated into two categories, surface water (directly collected from rivers, ponds, lakes, irrigation channels, etc.) and other unimproved sources (unprotected dug well, unprotected spring, water delivered by cart or tanker). Improved water sources are defined as those that, by nature of their construction, are protected from outside contamination, particularly faecal matter. Within improved sources,

N. Ronnie (✉) • N. Shah
Safety and Environmental Assurance Centre, Unilever R&D Bangalore,
64 Main Road, Whitefield, Bangalore 560066, India
e-mail: nirmala.ronnie@unilever.com; nimish.shah@unilever.com

P. McClure
Safety and Environmental Assurance Centre, Unilever R&D,
Colworth Science Park, Sharnbrook MK44 1LQ, UK
e-mail: peter.mcclure@unilever.com

K.J. Nath, V.P. Sharma (eds.), *Water and Sanitation in the New Millennium*,
DOI 10.1007/978-81-322-3745-7_6

<table>
<tr><td rowspan="10">World 6.9</td><td rowspan="8">Using Improved Sources 5.8</td><td rowspan="4">Piped 3.5</td><td rowspan="2">Safe 3.1 (2.9–3.2)</td><td>Low Sanitary Risk 2.8 (2.2–2.8)</td></tr>
<tr><td>Elevated Sanitary Risk** 0.33 (0.12–0.93)</td></tr>
<tr><td colspan="2">Unsafe * 0.11 (0.056–0.31)</td></tr>
<tr><td colspan="2">Unknown Safety 0.27</td></tr>
<tr><td rowspan="4">Other-Improved 2.3</td><td rowspan="2">Safe 1.3 (1.0–1.5)</td><td>Low Sanitary Risk 0.38 (0.31–0.42)</td></tr>
<tr><td>Elevated Sanitary Risk** 0.93 (0.62–1.2)</td></tr>
<tr><td colspan="2">Unsafe * 0.91 (0.69–1.3)</td></tr>
<tr><td colspan="2">Unknown Safety 0.11</td></tr>
<tr><td colspan="4">Using Unimproved Sources 0.78</td></tr>
<tr><td colspan="4">Unknown 0.30</td></tr>
</table>

* One billion people with improved sources that may be unsafe; ** 1.2 billion people with otherwise safe water with high sanitary risk

Fig. 6.1 World population by water contamination status and sanitary risk for 2010 (billions) (Sourced from Onda et al. [1])

the JMP differentiates between *piped to dwelling, plot or yard* and *other* sources such as public taps or standpipes, tube wells or boreholes, protected dug wells, protected springs and rainwater collection [2]. Most of the latter sources of water involve a significantly higher extent of handling of water in terms of collection, transport from the collection point and storage at home. It is important to recognize that on account of various technical, environmental, behavioural and management issues, linked to water access, distribution, transport and storage, even improved water sources continue to be at risk of microbiological contamination. Microbiological contamination can arise at many points in the catchment to consumer tap distribution chain [3] (Fig. 6.2).

Reports implicate distribution lines as an important cause of contaminated tap water. [4, 5] Such contamination may be attributable to (a) intermittent water supply leading to pressure changes, including vacuum which causes leakages and allows contaminants from outside into the supply, and (b) urban habitats consisting of organized and unorganized housing in close proximity that leads to unauthorized or repeated tapping of pipes that results in contamination.

Multilevel buildings requiring bulk storage and distribution of water, on account of factors such as faulty design, corrosion and improper maintenance, are vulnerable to microbiological contamination [6].

In a number of developing countries, '24×7' piped water remains elusive to a majority of urban and rural households. Households therefore need to resort to water storage for all domestic needs including that for drinking. Storage of water, particularly in compromised environmental settings, could lead to heightened risk of microbiological contamination [7, 8].

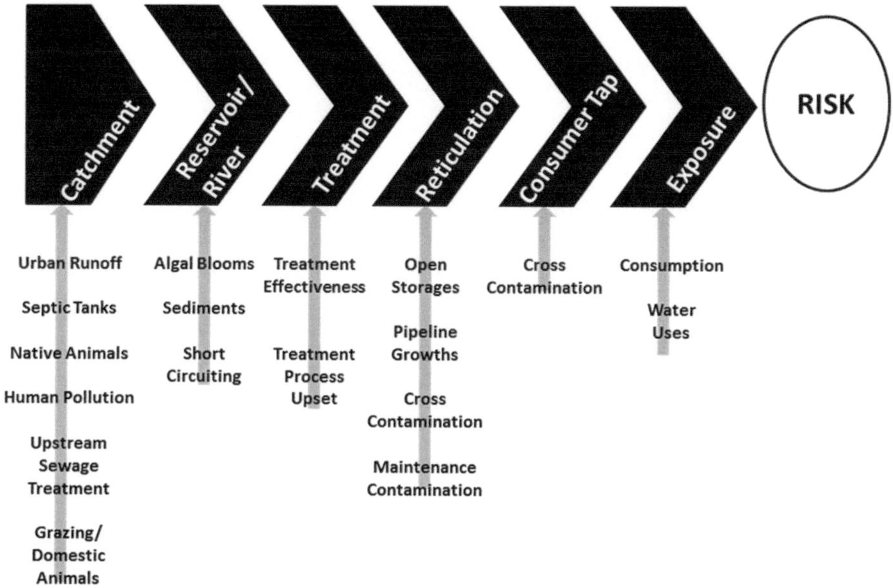

Fig. 6.2 Generic flow diagram for sources of microbiological risk in a drinking-water context (Adapted from Hunter et al. [3])

6.2 Microbial Pathogens in Drinking Water

One of the main sources of pathogenic microorganisms in drinking water is from human or animal excreta. Infectious diseases associated with drinking water are usually caused by exposure to three classes of enteric pathogens including bacteria, viruses and protozoan parasites that are transmitted by the 'faeco–oral' route. Each class of pathogen varies with regard to its physicochemical and biological characteristics, including its susceptibility to various treatment technologies (Table 6.1).

6.3 Importance of Household Water Treatment Systems (HWTS) to Address Infectious Disease Burden

Given the public health challenges around access to safe water and potential risks related to contamination during transport and storage, it is essential to consider complementing strategies to reduce exposure of the population to waterborne pathogens.

Drinking water treatment and safe storage at household level has been reported to offer significant public health benefits. HWTS are technologies, devices or methods applied for the purposes of treating water at the household level or at the point of use in community settings, like schools, healthcare facilities, etc.

Table 6.1 Pathogens associated with gastrointestinal diseases

Pathogen	Incidence and severity of disease	Relative infectivity[a]	Main reservoirs, human/zoonotic/both
Bacteria			
Campylobacter spp.	High	Moderate	Both
Escherichia coli pathogenic[b]	High	Low	Both
E. coli, enterohaemorrhagic	High	High	Both
Francisella tularensis	High	High	Both
Salmonella Typhi	High	Low	Human
Other salmonellae	High	Low	Both
Shigella spp.	High	High	Human
Vibrio cholerae	High	Low	Human
Viruses			
Adenoviruses	Moderate	High	Human
Astroviruses	Moderate	High	Human
Enteroviruses	High	High	Human
Hepatitis *A virus*	High	High	Human
Hepatitis E virus	High	High	Potentially both
Noroviruses	High	High	Potentially both
Rotaviruses	High	High	Potentially both
Sapoviruses	High	High	Potentially both
Protozoa			
Cryptosporidium hominis/parvum	High	High	Both
Cyclospora cayetanensis	High	High	Human
Entamoeba histolytica	High	High	Human
Giardia intestinalis	High	High	Both

[a]Infective dose, high 1–10^2 organisms, moderate 10^2–10^4 and low >10^4
[b]Pathogenic species includes enteropathogenic, enterotoxigenic, enteroinvasive, diffusely adherent and enteroaggregative
Adapted from WHO [9]

Meta-studies reviewing interventions at household levels through deployment of Household Water Treatment Systems (HWTS) have shown significant potential of contributing to reduction of disease burden [10–12]. WHO recognizes the use of HWTS as an effective measure for reducing disease burden linked to diarrhea [13].

In this context it is important to understand the channels through which households gain access to HWTS.

(a) Water treatment options distributed through donor funding which are in some form subsidized or funded through programmatic support – the primary intent of large-scale deployment of such HWTS technologies is to make an overall impact in the community and improve the public health statistics associated with waterborne disease burden.
(b) Water treatment options, specifically marketed to the household with the promise of providing safe water.

Relatively simple technologies that may not necessarily treat water to remove all forms of contaminants, and/or do not address a specific class comprehensively, have also been reported, across several intervention studies, to provide disease-burden reduction benefits across the population [12, 14]. The efficacy of different technologies developed for HWTS against waterborne microbiological hazards is summarized by WHO [13] and is based on published studies. The summary reports baseline reductions which are those expected in the field where unskilled personnel apply treatment to raw waters of varying quality and where facilities and practices do not allow for optimal treatment, and maximal reductions, which represent the highest level of performance from the technology used. The summary provides a useful reference point to understand which technologies work better for the different classes of pathogens likely to be present in source waters.

There are contexts where HWTS are actively marketed to consumers with a promise of safe treated water. In such situations, it is essential to ensure that HWTS offers comprehensive protection from all forms of waterborne microbiological pathogens, likely to be present. Further, it is important to ensure that critical elements relating to effectiveness and safety of the technology are addressed in order to ensure that consumers continuously access safe drinking water.

6.3.1 Risk-Based Approaches for Developing Microbiological Reduction Criteria

In terms of applying risk-based thinking globally, the US Environmental Protection Agency (US EPA) was the first regulatory body to adopt microbiological pathogen reduction targets for HWTS. Based on surveillance data and allowing for a certain minimum disease burden as a measure of acceptable risk, the US EPA [15] proposed that water, specifically of the type classified as unimproved (described above), could be rendered safe if the HWTS was to deliver a 6-$\log_{10}$ reduction of bacteria, 4-$\log_{10}$ reduction of viruses and 3-$\log_{10}$ reduction of parasitic cysts. These targets in practical terms were defined as follows:

(i) For bacteria:

'Based on the over 10^5/100 ml concentrations observed in highly polluted stream water and a target effluent concentration of less than 1/100 ml, a 6-log reduction is recommended'.

The 10^5/100 ml concentration refers to coliform bacteria.

(ii) For viruses:

'Assuming a target effluent level of less than one virus in 100 litres of water and a concentration of 10^4 enteric viruses in 100 litres of sewage-contaminated waters, the water purifier units should achieve at least 4 logs of virus removal'.

(iii) For protozoan cysts:

'From the lack of *Giardia* cases in systems where adequate filtration exists, a 3-log (99.9%) reduction requirement is considered to be conservative and to provide a comparable level of protection for water purifiers to a well-operated filtration treatment plant'.

The basis for target log reductions for viruses and protozoan cysts is described by Rose and Gerba [16], who stated that the geometric average for these agents should not exceed 10^{-3}/100 l following treatment, to achieve a 1:10,000 annual risk of infection. The risk assessment framework applied is described in Table 6.2. For the 6 $\log_{10}$ reduction of bacteria, a more general target was set for coliform bacteria, which is a group of indicator organisms, not all of which are pathogenic.

Recently, the WHO published a document 'Evaluating household water treatment options' [13], which outlined a separate tiered set of pathogen reduction criteria (for bacteria, viruses and protozoan parasites) which offer different levels of protection and described tiers such as interim, protective and highly protective. The top-tier standard of 'highly protective' represents those technologies that, if used correctly and consistently over an entire year, will limit drinking-water disease burden to 10^{-6} DALY (disability-adjusted life year) per person. According to WHO, this is a rather conservative health-based target, and, from health perspective, technologies that meet the pathogen reduction criteria applicable to the highly protective standard should be unequivocally recommended for use. The second tier, 'protective', was proposed to allow for a less stringent level of tolerable disease excess, yet it is still consistent with the goal of providing high-quality, safe water. The 'protective' target defines pathogen removals that achieve a health-based target of 10^{-4} DALY per person per year. In areas with a suspected high burden of waterborne disease, technologies that meet the log removal standards in the second tier would still result in significant health benefits. Recognizing that in the short term there may be challenges to develop affordable and scalable technologies that meet the highly protective or protective criteria, WHO proposed an interim standard – 'interim' target applies to those technologies that achieve 'protective' removal

Table 6.2 Risk assessment framework

Step	Aim
1. Problem formulation and hazard identification	To identify all possible hazards associated with drinking water that would have an adverse public health consequence, as well as their pathways from source(s) to consumer(s)
2. Exposure assessment	To determine the size and nature of the population exposed and the route, amount and duration of the exposure
3. Dose–response assessment	To characterize the relationship between exposure and the incidence of the health effect
4. Risk characterization	To integrate the information from exposure, dose–response and health interventions in order to estimate the magnitude of the public health problem and to evaluate variability and uncertainty

Source: Adapted from Hunter et al. [3]

targets for two classes of pathogens and have a proven impact on reducing diarrheal and waterborne infections.

On comparing US EPA and WHO, highly protective tier, it is evident that there are differences in the recommended treatment log reductions and this is attributed to the health targets set and also in the way in which the treatment log reduction is estimated. Considering health targets, an upper tolerable burden of disease limit of 10^{-6} is set by WHO and 1:10,000 annual risks is set by EPA. When the same input level of contamination for source water is used for the two approaches, the recommended log reduction to achieve the health target set by WHO is 100-fold higher than the target set by US EPA. This gives the impression that the health target set by WHO is more protective.

A good understanding of the data on frequency and levels of relevant pathogens in water is a key aspect of risk assessment, required for exposure assessment and determining the level of reduction required to deliver safe water. A recent study that used epidemiological data to carry out a quantitative risk assessment for an in-home water treatment device [17] concluded that the water used as source water for a HWTS being tested had lower contamination levels as compared to levels reported in waters from polluted urban environments. Indicators may vary in their ability to correlate with pathogens present and therefore may not reliably predict potential risks to human health, and this further emphasizes the importance of using reliable data that is relevant to water used as a source for HWTS.

At the country level, it is critical to appreciate the principles used by US EPA and WHO in arriving at the respective microbiological reduction criteria. Also, there is a need to consider generating surveillance data, especially for nonbacterial pathogens (such as viruses and parasites), and determine health targets for the country. These will serve as critical inputs to build log-reduction criteria that can be used as a basis for country standards for microbiological HWTS.

In addition to the comprehensive log-reduction criteria applicable to microbiological HWTS, it is important that standards consider the following additional elements for incorporation in technologies to ensure consumer safety:

HWTS and consumer safety – additional considerations:

- Consistency in removal effectiveness across the active lifetime of the technology
- A consumer alert mechanism to enable the consumer to change or service the device and/or components thereof on nearing exhaustion so that consumer can acquire the relevant materials and services in a timely manner
- Material of construction and/or active materials such as disinfectant should not leave residues or by-products in the finished water at levels that may be considered harmful for human consumption.

Clear labelling/communication that guides the consumer on the type of water that can be used with the HWTS, precautions and limitations, if any.

6.4 Conclusions

- In view of the challenges relating to water contamination, even when drawn from improved sources, deploying Household Water Treatment Systems that control microbiological pathogens in drinking water is emerging as a viable complementary approach for providing access to safe drinking water. It is important to understand the specific contexts relating to deployment of HWTS, whether it is through social channels aiming to reduce the overall disease burden in the population or specifically marketed to protect the household/family from waterborne diseases.
- Particularly important in the context of products directly marketed to households through advertisements and distributed through normal channels of trade and commerce are product standards, labelling and unambiguous consumer communication, which are critical for informed purchase decisions. Equally critical in such cases are the challenges in differentiation and communication related to products that offer varying degrees of pathogen reduction and therefore differential protection.
- Microbiological performance criteria for HWTS have been developed by US EPA and more recently by WHO. Both have used risk-based principles to develop the microbiological reduction criteria, with the WHO adopting a tiered approach, enabling inclusion of technologies relevant for social sectors yet offering a standard product.
- Locally relevant standards are needed to protect users; therefore, it is important that necessary country-level forums are built to look into data needs and formulate appropriate surveillance plans, both for determination of prevalence of pathogens and as basis of their epidemiological significance, and determine health targets which will form inputs to development of national standards for microbiological HWTS. In addition, specific consumer protection measures identified must be integrated in HWTS propositions.

To collect comprehensive and reliable data during pathogen surveillance in source waters, it is essential to develop a fit-for-purpose sampling plan, sufficient to reliably estimate the mean microbial levels, as well as magnitude of variability, and also extreme values. These data should include sampling during periods of greater vulnerability to contamination and likely higher risk (e.g. rainy season, impacts from known faecal contamination sources such as periodic sewage discharges and other site-specific factors that increase faecal contamination).

With increasing numbers of water treatment products entering the market, it is important to formulate national HWTS certification programmes which could certify technologies/products based on compliance to national performance standards. This in turn will require adoption of internationally accepted test protocols or development of testing and verification protocols relevant to local circumstances (e.g. availability of infrastructure).

Acknowledgements Contributions from SEAC colleagues, particularly Moira Johnston, Denis Kan-King-Yu and Niteen Sawant, towards development of risk-based opinion for HWTS formed the basis of this chapter and are greatly acknowledged.

References

1. Onda K, LoBuglio J, Bartram J (2012) Global access to safe water: accounting for water quality and the resulting impact on MDG progress. Int J Environ Res Public Health 9:880–894
2. WHO/UNICEF Joint Monitoring Programme for Water Supply and Sanitation. Progress on Sanitation and Drinking-Water (2010) Update. World Health Organization, Geneva, Switzerland, 2010
3. Hunter PR, Payment P, Ashbolt N, Bartram J (2003) Assessment of risk. In: Assessing microbial safety of drinking water: improving approaches and methods. OECD, WHO, Geneva, pp 79–109
4. Mahbub KR, Nahar A, Ahmed MM, Chakraborty A (2011) Quality analysis of Dhaka WASA drinking water: detection and biochemical characterization of the isolates. J Environ Sci Nat Resour 4(2):41–49
5. Pitkänen T, Miettinen IT, Nakari UM, Takkinen J, Nieminen K, Siitonen A et al (2008) Faecal contamination of a municipal drinking water distribution system in association with *Campylobacter jejuni* infections. J Water Health 6(3):365–376
6. WHO, Cunliffe D, Bartram J, Briand E, Chartier Y, Colbourne J, Drury D, Lee J, Schaefer B, Surman-Lee S (eds) (2012) Water safety in buildings. World Health Organization, Geneva
7. Taulo S, Wetlesen A, Abrahamsen R, Mkakosya R, Kululanga G (2008) Microbiological quality of water, associated management practices and risks at source, transport and storage points in a rural community of Lungwena. Malawi Afr J Microbiol Res 2:131–137
8. Wright J, Gundry S, Conroy R (2004) Household drinking water in developing countries: a systematic review of microbiological contamination between source and point-of-use. Tropical Med Int Health 9(1):106–117
9. WHO (2011) Guidelines for drinking water quality, 4th edn. Geneva, World Health Organization
10. Fewtrell L, Kaufmann RB, Kay D, Enanoria W, Haller L, Colford JM Jr (2005) Water, sanitation, and hygiene interventions to reduce diarrhoea in less developed countries: a systematic review and meta-analysis. Lancet Infect Dis 5:42–52
11. Clasen T, Roberts I, Rabie T, Schmidt W, Cairncross S (2006) Interventions to improve water quality for preventing diarrhea. Cochrane Database Syst Rev 19(3)
12. Waddington H, Snilstveit B, White H, Fewtrell L (2009) Water, sanitation and hygiene interventions to combat childhood diarrhea in developing countries. International initiative for impact evaluation. Synthetic Review 001
13. WHO (2011) Evaluating household water treatment options: health-based targets and microbiological performance specifications. World Health Organization, Geneva
14. Arnold BF, Colford JM (2007) Treating water with chlorine at point-of-use to improve water quality and reduce childhood diarrhea in developing countries: a systematic review and meta-analysis. Am J Trop Med Hyg 76:354–364
15. USEPA (1987) Guide standard and protocol for testing microbiological water purifiers. United States Environmental Protection Agency, Office of Drinking Water, Washington, DC
16. Rose JB, Gerba CP (1991) Use of risk assessment for development of microbial standards. Water Sci Technol 24:29–34
17. Enger KS, Nelson KL, Clasen T, Rose JB, Eisenberg JNS (2012) Linking quantitative microbial risk assessment and epidemiological data: informing safe drinking water trials in developing countries. Environ Sci Technol 46:5160–5167

Nirmala Ronnie is a Microbiology Risk Assessor at the Unilever Safety and Environmental Assurance Centre (SEAC) Bangalore, India. She leads the science area on water microbiology. The goal of her programme is to develop microbiological risk assessment approaches and tools for household water treatment. She leads a number of research collaborations with leading academic experts across the globe, for understanding pathogen physiology in different water matrices, development of improved methods for capture and quantification of pathogens and to study pathogen response towards different treatment technologies.

Peter McClure is currently the Food Safety section manager for Mondelez Europe, leading a team that provides food safety advice and guidance for manufacturing facilities in the region, and is part a global Food Safety group that contributes to development of food safety programmes rolled out across the globe. Prior to his current role, he was at Unilever for more than 20 years where was involved in product and process design, most recently as the science lead for microbiological safety. He is a member of the International Commission on Microbiological Specifications for Foods and also a member of the Advisory Committee on the Microbiological Safety of Food, in the UK. He is a visiting Professor at the University of Leeds where he lectures on microbiological safety and is coeditor of Foodborne Pathogens (Woodhead Publishing).

Nimish Shah is currently director of Safety and Environmental Assurance Centre (SEAC, Bangalore), Unilever R&D Laboratory, Bangalore, and has worked with Unilever R&D for more than two decades. He is SEAC programme director for the Strategic Science Group, a new pillar in R&D, and director of WASH (Water Sanitation and Hygiene).

He is a member of global R&D leadership teams, established and has driven R&D programmes and delivered breakthrough inventions, across foods, personal care, naturals, water purifiers and biotechnology businesses (14 patents). He leads the delivery of risk and impact assessment of innovations for consumer, occupational and environment safety and sustainability.

He leads SEAC's and Unilever's external agenda and industry-academia-regulatory interfaces and has driven external impact across food safety, WASH and sustainability. He established scientific partnerships with key opinion leaders/institutes. He is member of the Bureau of Indian Standards committees on drinking water and drinking water purification devices and ex-chair of the Standards Committee of the Water Quality Association, India. He pioneered setting standards for microbiological water treatment devices. He is member of the board of trustees of the International Life Sciences Institute, India, member of CII-FACE Food Safety Expert Committee and cochair of CII-HUL Initiative on Food Safety Sciences.

He has a PhD degree in microbiology (life sciences) and has also qualified the advanced course in food microbiology from the University of Surrey, UK.

7 Pesticides, Heavy Metals, and Fluoride Contamination of Groundwater Sources: Global and National Perspective

Prahlad K. Seth

7.1 Introduction

An essential requirement of a healthy life is safe drinking water; unfortunately around 1200 million people worldwide and around 125 million in India do not have access to same [1]. Waterborne diseases continue to be a major cause of morbidity and mortality in many parts of the developing world, and the most affected population is children. Supply of adequate quantities of drinking water to ever-increasing population continues to be a challenge in India and many developing countries. Surface waters, such as water bodies, rivers, ponds, and natural or man-made reservoirs, and groundwater are the main sources of drinking water in rural and urban settings. Water sources contain natural contaminants, particularly inorganic, as water flows through the geological strata, as well as contaminants from anthropogenic sources. Groundwater is a critical resource in India, as over 65% of water used for irrigation and 85% of water for human consumption are sourced from groundwater, particularly in rural areas. Generally, the groundwater is less vulnerable to pollution as compared to surface waters, yet groundwater contamination has been associated with severe diseases such as skin lesions, cancer, and other ailments arising from ingestion of water containing arsenic and dental and skeletal fluorosis from water containing fluoride in more than the permissible limits. Being geogenic, the groundwater chemical contamination is restricted to specific areas, but it affects large populations and is a significant public health problem in many parts of the world. WHO has provided guidelines for drinking water, and relevant agencies of most countries have prescribed the acceptable or permissible levels for chemical

P.K. Seth (✉)
Indian Institute for Toxicology Research (IITR), Sector G, Jankipuram, Kursi Road, Lucknow 226021, Uttar Pradesh, India
e-mail: prahladseth@gmail.com

K.J. Nath, V.P. Sharma (eds.), *Water and Sanitation in the New Millennium*,
DOI 10.1007/978-81-322-3745-7_7

and microbial contaminants in water which are revised from time to time. The Bureau of Indian Standards (BIS) has prescribed the acceptable levels of chemical and microbial contaminants which are also revised from time to time.

7.2 Anthropogenic Sources of Groundwater Pollution

In addition to geogenic sources, groundwater is reported to get contaminated from industrial wastes such as tanneries, chromium plating, pesticide, petroleum, etc. and agricultural chemicals like chlorinated pesticides and disinfecting agents. Groundwater also gets contaminated from spills and effluxes of hazardous waste sites. In India, very little data about the leachates from the hazardous waste sites is available, but a variety of hazardous waste is being dumped to unidentified and uncontrolled sites. Therefore, contamination of the groundwater from the leachates cannot be ruled out.

Groundwater status in India is being monitored by Central Ground Water Board (CGWB) through the State Ground Water Boards twice a year, and the data up to March 2012 is posted on their website (CGWB website). CGWB has also undertaken a study of contamination of groundwater in about 25 metropolis, and the data is available up to year 2008 (CPCB website). Besides this some universities and research institutions are also studying the levels of chemical contaminants in groundwater, but the studies are limited to few states or cities and are not regular. The presence of varying amounts of nitrate, fluoride, arsenic, selenium, iron chromium, chlorinated pesticides, residues and metabolites of disinfectants, residues of pharmaceutical agents, and other chemicals has been reported in groundwater [2]. New pockets of groundwater contamination from arsenic and fluoride are being reported. This chapter briefly deals with contamination of groundwater by pesticides, heavy metals, and fluoride.

7.2.1 Pesticides

Pesticides and insecticides are routinely used to contain pests and insects in agricultural practices. Their residues both in surface and groundwater are detectable worldwide [3]. The pesticides of concern are chlorinated pesticides which have low biodegradability and hence persist in the environment for long term. Their residues besides soils have been found in surface water mostly due to runoff from agricultural soils and discharge of waste water. Of significance is their presence in river sediments.

Pesticide's potential to contaminate water depends on several factors such as its solubility in water; factors influenced by soil, weather, season, as well as distance to water sources; and methods of their application. The runoff from pesticide-treated areas and their leaching through the soil into groundwater are one of the most common routes by which pesticide residues enter the water supplies.

Contamination of well water, located near the pesticide industries and agricultural fields, is quite feasable. More than 70 pesticide residues have been reported in groundwater from the United States [4]. Aldicarb and atrazine along with the soil fumigants were most frequently detected in groundwater [5]. From a study carried out in Pakistan, in 37 rural open wells located in Bahawalnagar, Muzaffargarh, and D.G Khan and Rajanpur districts of Punjab, the presence of six pesticides in water has been reported [6]. Varying quantities of pesticides both in surface and groundwater have been reported from different parts of India and have been reviewed. In most cases, the levels were within the permissible levels [7]. Although several pesticides have been banned in India and the use of pesticides like DDT, HCH, lindane, etc. has been restricted for agricultural purposes, their presence in water is still being detected.

A 2-year extensive study conducted by Indian Institute of Toxicology Research, Lucknow, India, in water and bed sediments of river Gomti at eight different sites reported the presence of persistent organochlorine pesticide (OCPs), namely, aldrin, dieldrin, endrin, HCB, HCH isomers, DDT isomers/metabolites, endosulfan isomers (alpha and beta), endosulfan sulfate, heptachlor and its metabolites, alpha-chlordane, gamma-chlordane, and methoxychlor [8]. The OCP residue levels in river water ranged between 2.16 and 567.49 ng/L, while the same was 0.92 and 813.59 ng/g in sediment. It was also observed that bed sediments are contaminated with lindane, endrin, heptachlor epoxides, and DDT. These may collectively contribute to sediment toxicity in the freshwater ecosystem of the river. The OCP concentrations in surface water and sediments in Tamiraparani River basin of south India ranged between 0.1–79.9 ng/L and 0.12–3,938.7 ng/g dry weight (dry weight) [9], and heptachlor, o,p′-DDE, dieldrin, o,p′-DDD, and mirex were the dominant OCPs in water samples indicating heterogenic nature of nonpoint source of pollution. The possibility of contamination of groundwater particularly in nearby dug wells is quite eminent.

Long-term, low-dose exposures to pesticides are increasingly being linked to adverse human health effects such as immunosuppression, hormone disruption, diminished intelligence, and other neurological problems, reproductive abnormalities, and cancer [10]. Nalini et al. [11] reported the presence of γ-HCH, malathion, and dieldrin. Interestingly enough, DDE, DDT, aldrin, ethion, methyl parathion, and endosulfan were undetectable in the samples collected from surface and groundwater of Kanpur district in northern India.

Jayasree and Vasudevan [12] in a study undertaken to detect the levels of organochlorines in the groundwaters of Thiruvallur district in Tamil Nadu reported high contamination with pp-DDT, op-DDT, endosulfan, HCH, and their derivatives. Also the presence of DDT, BHC, carbamate, endosulfan, etc. in groundwaters has been reported (Kumar and Shah @ www.indiawaterportal.org).

The OCPs being fat-soluble compounds are hard to excrete from the body. For these reasons, OCPs that bioaccumulate in the adipose tissue of mammalian species persist in soil and also get biomagnified through the food chain [13]. The food material cultivated on soil contaminated with pesticides or irrigated with pesticide-contaminated water often gets contaminated with pesticides, exposing population consuming such food material at risk.

7.2.2 Heavy Metals

Contamination of groundwater with metals of concern to health is predominantly geogenic among which arsenic is of great concern worldwide. Metals such as aluminum, cadmium, chromium lead, and mercury also cause grave concerns for human health. Most of these are heavily used in industry and can reach groundwater through excessive discharges in near wells. Seepage from industrial waste sites to the ground has also been a source of contamination. A study conducted in Ahmedabad in 2009 revealed the range of cadmium (Cd), arsenic (As), and lead (Pb) to be above the permissible standards of BIS and WHO. Alarming levels of hexavalent chromium ranging from 1.05 to 35.34 ppm have been reported in groundwater samples collected from a village of Kanpur Dehat [Personal communication Krishna Gopal, IITR]. In Kolkata, drinking water supply has been reported to be contaminated with lead, chromium, cadmium, and mercury [14]. The presence of heavy metals has been reported in the groundwaters of 40 districts in 13 Indian states – Andhra Pradesh, Assam, Bihar, Haryana, Himachal Pradesh, Karnataka, Madhya Pradesh, Orissa, Punjab, Rajasthan, Tamil Nadu, Uttar Pradesh, West Bengal [15], and Gurgaon and Mewat districts [16].

7.2.2.1 Manganese

Manganese which is required by the body for some essential biological functions is reported to lead to neurological disturbances in high quantities. However, in comparison with arsenic, there are not many reports of severe health ailments from exposure to manganese, and groundwater contamination is only in few selected mining and industrial areas. The groundwater contamination and associated adverse health effects of selected metals are described briefly here. For India, the Bureau of Indian Standards (BIS) has prescribed acceptable levels of metals for drinking water [17].

The presence of manganese (Mn) in drinking water is associated with neurological disorders, while lead (Pb), in addition to its effects on cognitive functions in children, affects other body functions as well. In a study conducted in the United States, manganese concentration was measured in home tap water and children's hairs among 362 children aged 6–13 years using groundwater. It was observed that higher levels of manganese in water and hair were significantly associated with lower IQ scores. There was a 6.2-point difference in IQ between children exposed to the lowest and highest levels of manganese in water. Manganese in water was more strongly associated with performance IQ than verbal IQ [18].

The problem of high manganese levels is not limited to any specific geographic region; in New England, 45% of wells for public use have manganese concentrations >30 μg/L [19]. Across the United States, roughly 5% of domestic household wells have alarmingly high concentrations of >300 μg/L [20]. Elevated manganese levels in groundwater are common in several countries, including Sweden [21], Vietnam [22], and Bangladesh [23].

7.2.2.2 Arsenic

Arsenic poisoning has become a public health problem in several countries, particularly Indian subcontinent (Bangladesh and Bengal), South America, and the Far East. This is mainly due to heavy dependence on arsenic-contaminated groundwater as source of drinking water. Like other metals, contamination of groundwater from arsenic has also been reported in other parts of the world, namely, Taiwan [24], Argentina, Chile, Hungary, Mexico, Northern China, and several parts of the United States [25, 26]. One of the biggest sources of arsenic toxicity is the geological composition of the aquifers that help in leaching of arsenic into the groundwater. The arsenic is dissolved via oxygenation that further facilitates the formation of soluble arsenate and arsenite [27, 28]. A study conducted by the School of Environmental Studies, Jadavpur University, West Bengal, that included 3417 villages and over 111 blocks, primarily within 12 districts of the West Bengal, namely, Koch Bihar, North and South Dinajpur, Maldha, Murshidabad, Nadia, Bardhaman, Kolkata, Hugli, Pargana adjoining river Bharathi, revealed that most of these regions have high arsenic levels that are unsafe for human consumption [29].

Interindividual variation has been suggested in arsenic metabolism [30], mainly linked to genetic polymorphism of enzymes participating in arsenic methylation. Monomethylarsonous acid (MMA^{III}) has been reported to be more cytotoxic and genotoxic than arsenate and arsenite [31]. Significant differences in genetic polymorphism have been observed among Asian (especially Japanese, Koreans, Chinese, Mongolians, Uygurs, Tibetans, Tamangs, Vietnamese, Tamils, and Sinhalese populations) and among non-Asian populations from Turkey, Ovambos, Ghana, and Xhosas and Caucasian American, Central European, Chilean, and African American [32].

More than 26 million people are exposed to arsenic through drinking water in West Bengal, India, as per an estimate, but fortunately, only 15–20% exhibit arsenic induced non-cancerous, precancerous, and cancerous skin lesions [33]. These differences in exposure and disease presentation among the same population can be attributed to individual's ability linked to genetic polymorphism to metabolize arsenic. Giri and his associates have substantiated the fact regarding the polymorphism-based susceptibility of individuals to arsenic poisoning [33].

Arsenic intoxication is reflected as early clinical symptoms that may include abdominal pain, nausea, vomiting, diarrhea, myalgia, and weakness with flushing of the skin. If clinically ignored, these symptoms may aggravate to numbness and tingling of the extremities, muscular cramping, and the appearance of an erythematous rash. Furthermore, dermal lesions; peripheral neuropathy; skin, lung, and bladder cancer; and peripheral vascular disease are common, following chronic exposure to arsenic-contaminated drinking water. The major dermatological signs following arsenic exposure are hyperpigmentation, melanokeratosis, hyperkeratosis, melanosis, spotted and diffused keratosis, leucomelanosis, dorsal keratosis, and peripheral vascular disease [34]. Besides cancer and skin diseases, consumption of water with high arsenic levels results in cardiovascular diseases, diabetes, and reproductive disorders. Diminished immunity has been reported among individuals ingesting low levels of arsenic through drinking water [35]. The possible mechanisms of arsenic poisoning have been elegantly described by several investigators and that helped in the better understanding of disease pathogenesis [36, 37].

7.2.2.3 Fluoride

Ingestion of water containing high levels of fluoride causes fluorosis, which has affected a vast population of the world including the Indian subcontinent, Africa, and the Far East. In the Southeast Asian region, fluorosis is common in Myanmar, Sri Lanka, and Northern Thailand [38]. One of the most common adverse effects of fluorosis is dental fluorosis that is characterized by brown mottling of the teeth. If unchecked, fluorosis may result in stiffened and brittle bones and joints, deformities in knee and hip bones, and ultimately crippling [39]. Skeletal fluorosis with crippling effects can occur when water has fluoride concentration above 1 part per million (ppm); however, in certain communities, skeletal fluorosis may also occur with only 0.7 part per million [40–43].

Like other contaminants, the geological and physiochemical characteristics of the water bodies, the pore size and acidity of the soil and rocks through which water percolates, the temperature, the action of other chemical elements, and the depth of wells are some critical factors that influence the natural concentration of fluoride in groundwater. Due to these factors, the fluoride concentration in groundwater varies between 1 ppm and more than 35 ppm. In some geographical locations such as Kenya and South Africa, levels of fluoride exceed up to 25 ppm. In India, concentrations as high as up to 38.5 ppm have been documented. According to the surveys undertaken by Central Ground Water Board, fluoride levels are above the permissible limits with levels as high as 48 mg/L in 19 states of the country. The states with these alarming levels of fluoride are Andhra Pradesh, Assam, Bihar, Chandigarh, Delhi, Gujarat, Haryana, Jammu and Kashmir, Jharkhand, Karnataka, Kerala, Madhya Pradesh, Maharashtra, Orissa, Punjab, Rajasthan, Tamil Nadu, Uttar Pradesh, and West Bengal [44]. The problem of fluorosis spans over urban and rural areas, being more alarming in rural setups. About 65% of Indian villages are facing the risk of fluorosis (Kumar and Shah, @ www.indiawaterportal.org). The 2011 WHO [45] guidelines for drinking-water quality suggest the optimal fluoride concentration in drinking water to be below 1 mg/liter (1ppm) in areas with a warm climate, but the same can be relaxed up to 1.2 mg/L in cooler climates.

As per the Indian Standard Drinking Water Specifications (BIS 2004), the maximum permissible limit for fluoride in drinking water is 1.5 ppm, and the highest desirable limit is restricted to 1.0 ppm. Despite the strict guidelines, roughly 20 million people are severely affected by fluorosis. Furthermore, 40 million are at risk of endemic fluorosis. Ingestion of water containing high levels of fluoride poses health problems depending on the levels of fluoride. Mild to severe fluorosis has been reported from Andhra Pradesh, Jammu and Kashmir, Delhi, Haryana, Gujarat, Karnataka, Kerala, Maharashtra, Punjab, Tamil Nadu, Rajasthan, and Uttar Pradesh. Fluorosis is common in Karbi, Anglong, and Nagaon districts of Assam, and these have been linked to high fluoride levels in groundwater that is the major source of drinking water in these areas [46]. Some parts of Uttar Pradesh have fluoride concentrations in groundwaters higher than the permissible limits (1.5 mg/l). Consequences of these high levels are reflected in preliminary reports, indicating severe health disorders in parts of the Kachnarwa region, located in the upper Panda river basin, Sonbhadra District in Uttar Pradesh [47]. The concentration of fluoride in the groundwater ranges from 0.483 to 6.7 mg/l. Among the areas studied, Rohiniyadamar, Madhuri, Neruiyadamar, Gobardaha, and Kunrwa seem to be the worst fluoride-affected villages of Uttar Pradesh. Majority of the people in these areas present clinical symptoms of dental and skeletal fluorosis that includes mottling of the teeth, deformation of ligaments, bending of the spinal column, and early aging.

The fluoride levels in Tamil Nadu vary from 0.1 to 2.8 mg/L in pre-monsoon season to 0.4–4.0 mg/Lin post-monsoon season [48]. Recently, high levels of fluoride in groundwater have been observed in Mettur region [49].

7.3 Conclusion

It is apparent that chemical contamination of groundwater sources with metals, pesticides, and fluorides is of global as well as of national concern since it is associated with a wide variety of health ailments ranging from simple intestinal disorders to reproductive, neurological, and immune disorders and cancer. Levels higher than permissible limits of fluoride have been reported from 19 states of India, and among metals arsenic is of great public concern. There are only few reports of groundwater contamination by pesticides mostly of organic chlorinated ones. Since groundwater is a precious resource for drinking and agricultural purposes, it should not only be monitored regularly and protected, but efforts should be made to develop novel remedial measures to keep this source alive.

References

1. Kunjappy R (2008) Progress and prospects on water: for a clean and healthy world with special focus on sanitation. World water week in Stockholm-2008
2. Stuart M, Lapworth D, Crane E, Hart A (2012) Review of risk from potential emerging contaminants in UK groundwater. Sci Total Environ 416:1–21
3. Asi MR, Hussain A, Muhmood ST (2008) Solid phase extraction of pesticide residues in water samples: DDT and its metabolites. Int J Environ Res 2(1):43–48
4. Thodal CE (2009) Monitoring for pesticides in groundwater and surface water in Nevada, 2008: U.S. Geological Survey Fact Sheet 2009–3093
5. Ritter WF (1990) Pesticide contamination of ground water in the United States-a review. J Environ Sci Health 25:1–29
6. Tariq M, Shahzad A, Ishtiaq H (2003) Pesticides in shallow groundwater of Bahawalnagar, Muzafargarh, D.G. Khan and Rajanpur districts of Punjab, Pakistan. Environ Int 30:471–479
7. Seth PK (2008) Chemical contamination of water: Indian scenario. In: Nath KJ, Sharma VP (eds) Safe water for community health. The National Academy of Sciences, Allahabad, pp 15–25

8. Malik A, Ojha P, Singh KP (2009) Levels and distribution of persistent organochlorine pesticide residues in water and sediments of Gomti River (India)—a tributary of the Ganges River. Environ Monit Assess 148:421–435
9. Kumarasamy P, Govindaraj S, Vignesh S, Rajendran RB, James RA (2012) Anthropogenic nexus on organochlorine pesticide pollution: a case study with Tamiraparani river basin, South India. Environ Monit Assess 184:3861–3873
10. Aktar W, Sengupta D, Chowdhury A (2009) Impact of pesticides use in agriculture: their benefits and hazards. Interdisc Toxicol 2(1):1–12
11. Nalini S, Ajit KS, Rashmi S (2004) Organochlorine and organophosphorous pesticide residue in ground water and surface waters of Kanpur, Uttar Pradesh, India. Environ Int 31:113–120
12. Jayasree R, Vasudevan N (2006) Organochlorine pesticide residues in groundwater of Thiruvallur district, India. Environ Monit Assess 128:209–215
13. Crinnion WJ (2009) Chlorinated pesticides: threats to health and importance of detection. Altern Med Rev 14:347–359
14. Chacraverti S (2011) The Kolkata metropolitan area: probing for four toxic metals in drinking water. Project report (Project supported by EEJP) Society for Direct Initiative for Social and Health Action (DISHA)
15. Kaur R, Rani R (2006) Spatial characterization & prioritization of heavy metal contaminated soil-water resources in Peri-urban areas of National Capital Territory (NCT), Delhi. Environ Monit Assess 123:1–3
16. Kaur R, Minhas PS, Jain PC, Singh P, Dubey DS (2008) Geo-spatial analysis of land – water resource degradation in two economically contrasting agricultural regions adjoining national capital territory (Delhi). Environ Monit Assess. doi:10.1007/s10661-008-0378-3
17. Bureau of Indian Standards 10500–2004 Specifications for drinking water. Bureau of Indian Standards, New Delhi
18. Bouchard MF, Sauvé S, Barbeau B, Legrand M, Brodeur M-È et al (2010) Intellectual impairment in school-age children exposed to manganese from drinking water. Environ Health Perspect 119(1):138–143
19. Groschen GE, Arnold TL, Morrow WS, Warner KL (2009) Occurrence and distribution of iron, manganese, and selected trace elements in ground water in the glacial aquifer system of the Northern United States. U.S. geological survey scientific investigations report 2009–5006
20. U.S. Geological Survey (2009) National water-quality assessment program. Quality of water from domestic wells in the United States. Available http://water.usgs.gov/nawqa/studies/domestic_wells/table2.html
21. Ljung K, Vahter M (2007) Time to re-evaluate the guideline value for manganese in drinking water? Environ Health Perspect 115:1533–1538
22. Agusa T, Kunito T, Fujihara J, Kubota R, Minh TB, Kim Trang PT et al (2006) Contamination by arsenic and other trace elements in tube-well water and its risk assessment to humans in Hanoi. Vietnam Environ Pollut 139:95–106
23. Wasserman GA, Liu X, Parvez F, Ahsan H, Levy D, Factor-Litvak P et al (2006) Water manganese exposure and children's intellectual function in Araihazar, Bangladesh. Environ Health Perspect 114:124–129
24. Lamm SH, Engel A, Penn CA, Chen R, Feinleib M (2006) Arsenic cancer risk confounder in southwest Taiwan data set. Environ Health Perspect 114:1077–1082
25. Frost FJ, Muller T, Petersen HV, Thomson B, Tollestrup K (2003) Identifying US populations for the study of health effects related to drinking water arsenic. J Expo Anal Environ Epidemiol 13:231–239
26. Peters SC (2008) Arsenic in groundwaters in the Northern Appalachian Mountain belt: a review of patterns and processes. J Contam Hydrol 99:8–21
27. Kevin Henke (2009). Arsenic: environmental chemistry, health threats and waste treatment. Wiley, Hoboken, p 588. ISBN:978-0-470-02758-5

28. Akter A, Ali MH (2011) Arsenic contamination in groundwater and its proposed remedial measures. Int J Environ Sci Tech 8(2):433–443
29. Jadavpur University (2006) Report on ground water arsenic contamination in West Bengal, India – 20 years study. http://www.soesju.org
30. Vahter M (2000) Genetic polymorphism in the biotransformation of inorganic and its role in toxicity. Toxicol Lett 112–113:209–217
31. Petric JS, Ayala-Fierro F, Cullen WR, Carter DE, Vasken AH (2000) Monomethylarsonous acid (MMA^{III}) is more toxic than arsenite in Chang human hepatocytes. Toxicol Appl Pharmacol 163:203–207
32. Takeshita H, Fujihara J, Takatsuka H, Agusa T, Yasuda T, Kunito T (2009) Diversity of glutathione S-transferase omega 1 (A140D) and 2 (N142D) gene polymorphism in worldwide populations. Clin Exp Pharmacol Physiol 36:283–286
33. Banerjee N, Nandy S, Kearns JK, Bandyopadhyay AK, Das JK, Majumder P et al (2011) Polymorphisms in the TNF-alpha and IL10 gene promoters and risk of arsenic-induced skin lesions and other nondermatological health effects. Toxicol Sci 121:132–139
34. Fawell J, Nieuwenhuijsen MJ (2003) Contaminants in drinking water: environmental pollution and health. Br Med Bull 68:199–208
35. Kozul CD, Ely KH, Enelow RI, Hamilton JW (2009) Low-dose arsenic compromises the immune response to influenza: a infection in vivo. Environ Health Perspect 117:1441–1447
36. Mazumder GD, Dasgupta UD (2011) Chronic arsenic toxicity: studies in West Bengal, India. Kaohsiung J Med Sci 27:360–370
37. Singh AP, Goel RK, Kaur T (2011) Mechanisms pertaining to arsenic toxicity. Toxicol Int 18:87–93
38. World Health Organization (2010) Drinking water quality in the South-East Asia region. SEA-EH-567
39. Sharma S, Ramani J, Bhalodia J, Thakkar K (2012) Fluoride and fluorosis in context to Gujarat state of India: a review. Res J Pharm Biol Chem Sci 3:85–94
40. Susheela AK (1993) Prevention and control of fluorosis. Vol. 1 Health aspects. Published by Ministry of Rural Development (RGNDWMA), Government of India
41. Susheela AK, Kumar A, Bhatnagar M, Bahadur R (1993) Prevalence of endemic fluorosis with gastrointestinal manifestations in people living in some north Indian villages. Fluoride 26:97–104
42. Susheela AK (1995) Epidemiological studies of health risks from drinking water naturally contaminated with fluoride. Proc Rome Symp 233:123–134
43. Janardhana RN, Sangita D, Das K (2011) Emerging challenge: fluoride contamination in groundwater in Agra District, Uttar Pradesh. Asian J Exp Biol Sci 2:131–134
44. INSA (2011) Hazardous metals and minerals pollution in India: sources, toxicity and management. A position paper. Indian National Science Academy New Delhi
45. World Health Organization (2011) Guidelines for drinking water quality, 4th edn. World Health Organization, Geneva
46. Das B, Talukdar J, Sarma S, Gohain B, Dutta RK, Das HB et al (2003) Fluoride and other inorganic constituents in groundwater of Guwahati, Assam, India. Curr Sci 85:657–661
47. Raju NJ, Dey S, Das K (2009) Fluoride contamination in groundwaters of Sonbhadra District, Uttar Pradesh, India. Curr Sci 96:979–985
48. Sendesh K, Ramasubramanian V (2011) Assessment of fluoride contamination in groundwater using GIS, Dharmapuri District, Tamil Nadu, India. Int J Eng Sci Technol 3:1077–1085
49. Srinivasamoorthy K, Vijayaraghavan V, Vasanthavigar M, Sarma S, Chidambaram S, Anandhan P et al (2012) Assessment of groundwater quality with special emphasis on fluoride contamination in crystalline bed rock aquifers of Mettur region, Tamil Nadu, India. Arab J Geosci 5:83–94

Prof. Prahlad K. Seth is senior advisor at Biotech Park and its founding CEO. He is a former director of the Indian Institute of Toxicology Research, Lucknow. He is adjunct professor of toxicology at Jamia Hamdard, New Delhi, and honorary professor of biochemistry at Lucknow University, Lucknow.

Prof. Seth, an outstanding scientist, has made significant contributions to toxicology, neurosciences, molecular biology and biotechnology during the last four decades. He has played an important role in the growth of toxicology in India and identifying mechanisms of toxicity of pesticides, metals and monomers. He has made significant contributions to the National Drinking Water Mission by developing a portable water analysis kit and a mobile laboratory van for analysis of chemical and biological contaminants.

Prof. Seth is an elected fellow of the Collegium Internationale Neuro-Psychopharmacologicum, Indian National Science Academy, National Academy of Medical Sciences, the National Acdemy of Sciences India and other bodies. He has been conferred several awards including the Prof. B.K. Anand Oration of the National Academy of Medical Sciences, Vigyan Bhushan Award of the UP Council of Science and Technology and P.C. Ray Memorial Award of the Indian Science Congress Association. He has published over 240 research papers and contributed more than 50 review articles/book chapters and holds 10 patents

Geogenic Contamination and Technologies for Safe Drinking Water Supply

8

Narayan C. Ghosh

8.1 Introduction

Geogenic contaminants in groundwater are those which originate from the rock material by weathering and deposit in the aquifer and enter into aqueous phase by the processes of natural soil/rock-water interaction. Among the geogenic contaminants, arsenic and fluoride are the most widespread affecting health of millions of people the world over by the intake of excessive amounts beyond the permissible limit. The guideline values of the World Health Organization (WHO) for drinking water are 10 μg/l for arsenic and 1.5 mg/l for fluoride, and presently India also follows the same guidelines. These geogenic substances are mobilized from aquifer under certain geochemical and geological conditions. Despite these facts, arsenic- and fluoride-contaminated aquifers are continued to have tapped for drinking and irrigation water supplies because of the growing scarcity of water resources, and groundwater is mostly preferred in India as reliable source for drinking and irrigation water in rural areas.

Exposure to arsenic can cause variety of disorders, ranging from changes in skin pigmentation, hyperkeratosis and cardiovascular problems to cancer [10]. While small amounts of fluoride provide protection against caries and strengthening of bones, elevated concentration in water can lead to irreversible fluorosis [3].

India is one of the large-scale victimized countries of arsenic and fluoride contamination in groundwater. A number of district covering approximately 88,688 km^2 and 50 million people in seven states, namely, West Bengal, Bihar, Uttar Pradesh, Jharkhand, Assam, Manipur and Chhattisgarh, have been reported exposed to groundwater arsenic contamination above 50 μg/l [16], while about 66.62 million people, out of which are 6 million children below the age of 14 years from 19 states, namely, Andhra Pradesh, Assam, Chhattisgarh, Delhi, Gujarat, Haryana, J & K,

N.C. Ghosh (✉)
Ground Water Hydrology Division, National Institute of Hydrology, Roorkee 247 667, India
e-mail: ncg_1959@rediffmail.com

K.J. Nath, V.P. Sharma (eds.), *Water and Sanitation in the New Millennium*,
DOI 10.1007/978-81-322-3745-7_8

Jharkhand, Karnataka, Kerala, Madhya Pradesh, Maharashtra, Orissa, Punjab, Rajasthan, Tamil Nadu, Uttar Pradesh and West Bengal, have been reported exposed to fluoride contamination in groundwater above 1.5 mg/l [26]. With every additional survey, new arsenic- and fluoride-affected villages and people suffering from arsenic- and fluoride-related diseases are being reported [16].

Although the exact sources and geochemical processes are yet to be established, however, the cause has been recognized to be of geogenic origin, and the contaminants are released from soil under conditions conducive to dissolution from solid phase on soil grains to liquid phase in water. Whether the understanding and technological options available are adequate to resolve the issues or there are further needs to undertake, more investigations to strengthen understanding of geochemical processes to mitigate and remediate geogenic contaminants in groundwater are some of the concerns that need to be addressed for attaining sustainability in supply of safe groundwater in the affected areas.

It is in those contexts, the chapter is focused to give an insight on sources of geogenic contaminants, probable causes of attribution, geochemical processes that dominate mobilization in the aquifer systems, etc. to help suggest some possible conservation and remedial measures to restore back dependability of using groundwater resources with less health risks.

The detailed illustrations on clinical health risks, social aspects, scope of removal technologies developed, scale-up of the problem in different areas, geochemistry part and behaviours of the contaminants that coexist with other chemical constituents, which are available in many literatures, are kept beyond the purview of this paper.

8.2 Arsenic Menace in India

Arsenic contamination in groundwater has been reported, mostly in areas formed by recent alluvial sediments, describing Holocene aquifers (< 12 thousand years of age) of the Ganga-Brahmaputra plains. Almost all the identified arsenic-affected areas, in the Gangetic plains, are in a linear track on both sides of the River Ganga in UP, Bihar, and Jharkhand or the River Bhagirathi in West Bengal (Fig. 8.1), except areas in Chhattisgarh and three districts of Bihar, namely, Darbhanga, Purnea and Kishanganj. The areas in Assam and Manipur are in the flood plains of the Brahmaputra and Barak, respectively [16]. Ironically, all the arsenic-affected areas have the river routes originated from the Himalayan region. Whether the source material has any bearing to the outcrops or not is a matter of research, however, over the years, continued exploitation of groundwater has made the problem of arsenic contamination more complicated, to a large extent at both local and regional scale, by a number of unknown factors.

Arsenic in groundwater in the Holocene aquifers, which are of Quaternary age and comprise of micaceous sand, silt and clay derived from the Himalayas, is believed to be released from soil, under conditions, conducive to dissolution of arsenic from solid phase on soil grains to liquid phase in water clay [4, 12].

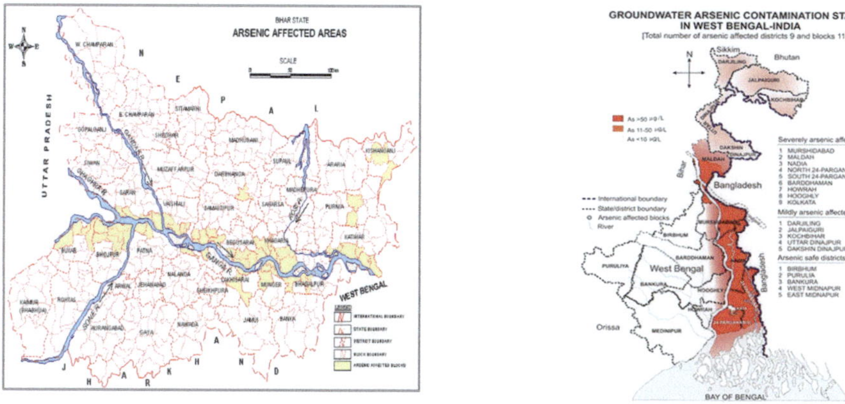

Fig. 8.1 Arsenic-contaminated areas along the River Ganga in Uttar Pradesh, Bihar and along the river Bhagirathi in West Bengal [16]

8.2.1 Hypothesis on Origin of Arsenic

A numerous speculations about the primary source of arsenic in the Bengal basin are available in literature. Several investigators have given varied opinions on origin of arsenic; however, commonality of those opinions is natural phenomena attributed by anthropogenic interferences [1, 2, 17]. The prominent hypotheses are:

(i) Arsenic has been transported by the River Ganges and its tributaries from the Gondwana coal seams in the Rajmahal trap area located at the west of the basin [19].
(ii) Arsenic has been transported by the North Bengal tributaries of Bhagirathi and Padma near the Gorubathan base-metal deposits in the Eastern Himalayas [17].
(iii) Arsenic has been transported with the fluvial sediments from the Himalayas [12] and deposited in the Holocene aquifer. This is the most accepted hypothesis at present.

The physical processes associated with the most commonly accepted hypothesis as explained by many investigators [15, 23, 24] are:

The attribution of arsenic in groundwater is from the subsurface sediments. These subsurface sediments originated from the mountains in the upstream river catchment had been deposited thousands of years back, and the deposited material formed the aquifer. Mountain erosion led to a release of rock-forming minerals and arsenic into the hydrosphere. Eroded iron turned to rust, iron (hydroxide) [(FeO(OH)] and formed particles as well as coatings on the surface of particles as such silt and sand. The FeO(OH) was capable to scavenge dissolved arsenic from water and bound it to its surface. Suspended particles with FeO(OH) coatings and adsorbed arsenic were washed into rivers and transported downstream. Arsenic bound to suspended solids was thus brought to rivers' deltas and deposited in the soils with the settling sediments. For thousands of years, deposits of river sediments have created the soil layers (sediments) that formed the delta as it is known today.

8.2.1.1 Hypothesis on Occurrence and Mobilization of Arsenic in Groundwater

Numerous literatures [13, 15, 20, 23] explaining occurrence of arsenic in groundwater, particularly in the alluvial aquifers of the Ganges delta, are available. Based on arsenic geochemistry, three hypotheses describing probable mechanisms of As mobilization in groundwater specially, with reference to Holocene aquifers like in West Bengal and Bangladesh, have been suggested [5]. These are:

(i) Mobilization of arsenic due to the oxidation of **As**-bearing pyrite minerals: Insoluble **As**-bearing minerals, such as arsenopyrite (FeAsS), are rapidly oxidized when exposed to atmosphere, releasing soluble As^{3+}, sulphate (SO_4^{2-}) and ferrous iron (Fe^{2+}). The dissolution of these **As**-containing minerals is highly dependent on the availability of oxygen and the rate of oxidation of sulphide. The released As^{3+} is partially oxidized to As^{5+} by microbial-mediated reactions. The chemical reaction is given by:

$$\mathrm{FeAsS} + 13\,\mathrm{Fe}^{3+} + 8\,\mathrm{H_2O} \rightarrow 14\,\mathrm{Fe}^{2+} + \mathrm{SO_4}^{2} + 13\,\mathrm{H}^{+} + \mathrm{H_3AsO_4}\ (\mathrm{aq.})$$

(ii) Dissolution of **As**-rich iron oxyhydroxides (FeOOH) due to the onset of reducing conditions in the subsurface: Under oxidizing conditions and in the presence of **Fe**, inorganic species of **As** are predominantly retained in the solid phase through interaction with FeO(OH) coatings on soil particles. The onset of reducing conditions in such environments can lead to the dissolution of FeO(OH) coatings. Fermentation of peat in the subsurface releases organic molecules (e.g. acetate) to drive reducing dissolution of FeO(OH), resulting in release of Fe^{2+}, As^{3+} and As^{5+} present on such coatings. The chemical reaction is given by:

$$8\,\mathrm{FeOOH} - \mathrm{As}_{(s)} + \mathrm{CH_3COOH} + 14\,\mathrm{H_2CO_3} \rightarrow 8\,\mathrm{Fe}^{2+} + \mathrm{As}_{(d)} + 16\,\mathrm{HCO_3}^{-} + 12\,\mathrm{H_2O}$$

where $As_{(s)}$ is sorbed As, and $As_{(d)}$ is dissolved As.

(iii) Release of **As** sorbed to aquifer minerals by competitive exchange with phosphate ($H_2PO_4^-$) ions that migrate into aquifers from the application of fertilizers to subsurface soil.

The second mechanism involving dissolution of FeO(OH) under reducing conditions is considered to be the most probable reason for excessive accumulation of **As** in groundwater. Elaborating further to this hypothesis linking it to the most commonly accepted theory of origin of arsenic, the cause of As in groundwater has been explained as follows:

The accepted geochemical theory is anxious dissolution of FeO(OH) and release of previously absorbed arsenic. The arsenic remains fixed in the sediments as long as groundwater contains sufficient dissolved oxygen. Arsenic is released when these come in contact with the oxygen-depleted groundwater [12, 14, 15, 23]. During the inundation periods, high loads of river sediments (suspended particles) are frequently covering the topsoil layers including vegetation. This process results in the entrapment and subsequent burial of natural organic matter (rotten plant, peat) in the sediment structure. Organic matter can serve as substrate (food) for microorganism to thrive on. These microorganisms consume dissolved oxygen to degrade organic material, thereby leading to oxygen depletion in the groundwater (anoxic conditions). Under anoxic condition, some microorganisms can use FeO(OH) as a source of energy instead of oxygen. Degradation of solid FeO(OH) particles releases arsenic formerly attached firmly to the particle surface.

8.2.2 Present Status of Arsenic Problem

It is now generally accepted that the source is of geological origin and percolation of fertilizer residues might have played a modifying role in its further exaggeration. Identification of parental rocks or outcrops including their sources, routes, transport, speciation and occurrence in Holocene aquifers along fluvial tracks of the Ganga-Brahmaputra-Barak valley and in scattered places, adjoining to it, in their basins is yet to be studied comprehensively. The question of the possible role of excessive withdrawal of groundwater for its triggering, however, has continued to have divided opinions. Whether the processes of physicochemical transformation were influenced by excessive groundwater exploitation or there were other coupled actions of a number of hydrogeological and geo-environmental disturbances over the periods is yet to be established.

Since arsenic menace was first surfaced in year 1984, substantial works have been carried out mostly towards enrichment of knowledge and understanding, which can be categorized as follows:

(i) The source and cause of arsenic contamination in groundwater
(ii) Extent and magnitude of scale-up
(iii) Mechanism of dissolution of arsenic from soil phase to aqueous phase
(iv) Impact on people health: diagnosis of sickness and symptoms
(v) Development of technologies for removal of arsenic from extracted groundwater
(vi) Analytical techniques for detecting arsenic in groundwater

The understanding and knowledge base accomplished from R & D activities are not adequate to resolve the problem completely. The counteractive and precautionary measures initiated by the government are not sufficient to provide sustainable solution to meet the water demands of the rural populace. Numerous investigations have come out with a number of findings, alternatives and propositions, which vary from identification of shortfalls to success stories. In this context, a detailed compilation made by NIH and CGWB [16] is referred. The present state of affairs of the problem in many states of India demands a systematic translation of success stories of one place/region to another, overcoming the shortfalls by conceiving R & D studies in areas wherever they are deemed fit.

Although the calamity of groundwater arsenic contamination in the other states is not as old as and as serious as it is in West Bengal, however, scaling up and surfacing of groundwater arsenic with every additional survey in a number of districts pose a serious threat towards further exploitation and uses of those contaminated aquifers and also to the people using the contaminated groundwater in different forms. Studies carried out and action taken so far in other states to understand the problem-resolving issues, counteractive measures, etc. are meagre in comparison to West Bengal, while characteristics and features of the problem, geological formations and causes of the problem are largely similar. Thus, the experiences and knowledge base acquired so far from the West Bengal could help evolve a framework of activities and sustainable mitigation strategies for other states as well.

8.3 Fluoride Menace in India

Fluoride contamination in groundwater in 19 out of 35 states and UTs in India and its intake with drinking water in excess to the permissible limit of 1.5 mg/l has emerged as more serious problem than arsenic from an endemic disease known as 'fluorosis' [7]. Its occurrence is quite widespread that varies from one hydrogeological and geological setting to another. About 70–100% districts in Andhra Pradesh, Gujarat and Rajasthan; 40–70% districts in Bihar, Delhi, Haryana, Jharkhand, Karnataka, Maharashtra, Madhya Pradesh, Orrisa, Tamil Nadu and Uttar Pradesh; and 10–40% districts in Assam, Jammu and Kashmir, Kerela, Chattisgarh and West Bengal have been reported affected by fluoride contamination in groundwater, while the endemicity for the rest of the states is not known. The fluoride level in the affected states ranges from 0.1 mg/l up to a maximum of 29 mg/l. Fig. 8.2 [26] depicts percentage districts and range of fluoride concentration detected in different states of India. The occurrence mainly in top aquifer system, i.e. in shallow groundwater zone, has been reported due to the result of the geochemical processes of the source rock material. Unlike arsenic occurrence in groundwater in a linear track represented by fluvial alluvium sediments, fluoride contamination is widespread in different states represented by varied hydrogeological formations. The only commonality with arsenic problem is that both have geogenic source and are attributed by the geochemical processes of the source rock-water interaction. The hydrogeological formations of the fluoride-contaminated aquifer are mainly basalt; crystalline rocks, viz. granites, gneisses and schists; and Precambrian sedimentary, viz. consolidated sandstones, shales, limestones, Gondwana sedimentary, etc.

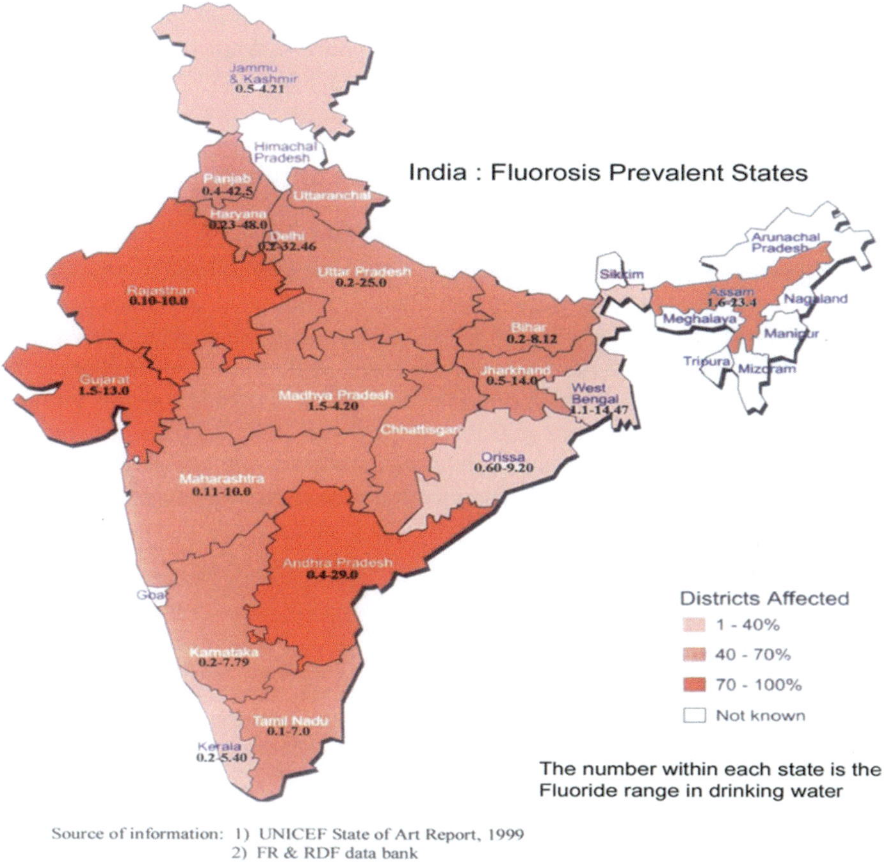

Fig. 8.2 Fluoride-contaminated states in India and percent districts affected in each state and fluoride range detected in drinking water [26]

8.3.1 Fluoride-Containing Rocks

The fluoride content of groundwater varies greatly depending on the geological settings and rock types. The most common fluorine-bearing minerals are fluorite, apatite and micas. Fluoride problems are predominant in places where these minerals are most abundant in host rocks. Groundwater from crystalline rocks, especially granites (deficient in calcium), are particularly sensitive to relative high fluoride concentrations. Such rocks are found especially in Precambrian basement areas. Fluorine transport in the aqueous solutions is mainly controlled by the solubility of CaF_2 [6]. Sedimentary rocks also contain fluorine concentration. In carbonate sedimentary rocks, the fluorine is present as fluorite. Metamorphic rocks also contain fluorine concentration.

The geological formations represented by the states affected by fluoride contamination in groundwater possess similar rock types. The occurrence of fluoride in groundwater is not a phenomenal but a continued geochemical process of rock-water interaction whose triggering effect could notice through large-scale laboratory and clinical detection.

8.3.2 Factors Affecting Natural Fluoride Concentration

The ultimate fluoride concentration in groundwater largely depends on its reaction time with aquifer materials. Groundwater can have high fluoride concentration if it has longer residence time in the aquifers. Such groundwaters are associated with deep aquifer systems and a slow groundwater movement. Shallow aquifers which contain recently infiltrated rainwater usually have low fluoride concentration.

Arid regions are prone to high fluoride concentrations because of slow groundwater movement and thereby long residence time with rock materials. The fluoride contents in water may increase during high evaporation if solution remains in equilibrium condition with calcite and alkalinity. Fluoride increase is less pronounced in humid tropics because of high rainfall and their diluting effect on groundwater chemical composition.

8.3.3 Fluoride Removal Technologies

Defluoridation techniques, currently in practices, can broadly be divided in three categories according to the main removal mechanism:

- Chemical additive methods
- Contact precipitation
- Adsorption/ion exchange methods

Based on the above defluoridation mechanisms, a number of devices have been developed in which the Nalgonda technique (named after the village in Andhra Pradesh where the method was pioneered) which follows the mechanism of chemical additive method and uses alum (hydrate aluminium salts) as coagulant has acquired popularity. The other devices, which are based on adsorbent/ion exchange, such as activated alumina (Al_2O_3), activated charcoal, or ion exchange resins, are used for defluoridation of contaminated groundwater for both community and household uses.

8.4 Groundwater Usages Scenarios in India

In India, groundwater meets nearly 85% of rural domestic water needs, 50% of urban and industrial water needs and about 60% of irrigation requirements [8, 25]. Usages of groundwater in India have increased at a very rapid pace by the advent of tube wells as the groundwater extraction structure. The data of the Minor Irrigation

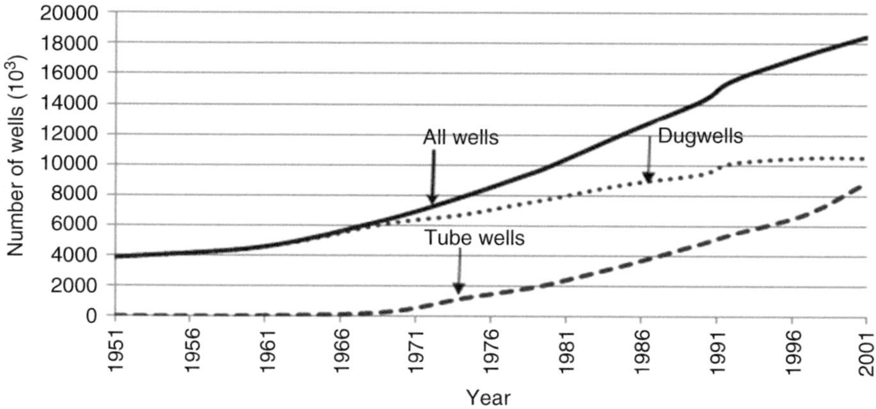

Fig. 8.3 Growth of groundwater irrigation structures in India (1951–2001) (Source: Graph prepared using data of Minor Irrigation Census, 2001 & data compiled by Singh & Singh, 2002)

Census conducted in 2001 together with the data compiled by Singh & Singh [22] shows (Fig. 8.3) enormous growth of groundwater structures. The number was around 18.5 million in 2001. Many people predicted [21] that by 2009, the number of groundwater irrigation structures might have gone up to 27 million. With such a huge number of groundwater abstraction structures and nearly 61% of status of groundwater development [8], India is placed now the largest groundwater user in the world [21].

Groundwater is mostly preferred because (1) rural people have a common notation that groundwater is less risk-free from pollution than surface sources of water, and (2) it is ubiquitous and can be drawn on demand wherever and whenever required. These general beliefs together with poor public irrigation and drinking water delivery system, new pump technologics, flexibility and timeliness of groundwater supply, government subsidy on electricity in the rural areas and lack of groundwater regulation legislation have given rise to preferential growth of groundwater uses in India. There is no rationale to believe that the growth of groundwater withdrawal structures and uses of groundwater in India are going to slow down in the future, unless otherwise controlled by enforcing legislation; rather, it will continue to rise because of growing concern on water quality, socio-economic improvement and sociocultural dimensions of the rural sector.

The rise of tube well extraction structures and their uses to large exploration depth, in many situations, have led to excessive withdrawal and overexploitation of aquifer than the annual groundwater recharge to it. Consequential effects of which have emerged in the form of persistent groundwater level depletion in many areas and triggering of contaminants earlier attached to the soil and rock formations to the water by the natural geochemical processes of soil-water interaction. Depletion of groundwater levels, on the one hand, provoking private players to dig tube wells at larger depth, and water in the deep aquifers, on the other hand, because of large contact times with the source materials, poses a serious threat of fluoridation.

In the case of arsenic, the intermediate aquifer depth that ranges between 20 m and 80 m has been reported contaminated, and its mobilization to the deep aquifer depends on its exploitation and movement direction of groundwater flow. In areas where deep aquifer is separated by an impervious layer from the overlain arsenic-contaminated zone, the impervious layer acts as barrier between the two layers.

8.5 Conservation, Mitigation and Remedy to Fluoride and Arsenic Contamination

The foremost task towards combating geogenic source of groundwater contamination should be to prepare risk maps delineating areas with elevated arsenic and fluoride concentrations in groundwater together with aquifer map. Most of the maps published represent detected limit of point measurement without referring to the spatial variation of the concentration. These resulted into a misinterpretation and varied opinions on the physical processes of attribution. Aquifer mapping representing geological formations together with the risk maps of contaminants can help understand and develop aquifer management plan to combat geogenic source of groundwater contamination.

8.5.1 Conservation of Arsenic Contamination

Arsenic contamination in groundwater has far-reaching consequences including its ingestion through food chain. Food is the second largest contributor to arsenic intake by people after direct ingestion of arsenic-contaminated water [16]. The concern, therefore, cannot be directed towards not only supply of arsenic safe drinking water but also supply of irrigation water. Arsenic removal technologies, which only consider ex situ treatment, cannot be an affordable task for supplying irrigation water. It is necessary to devise alternate methods for supply of both drinking and irrigation water. Ex situ arsenic treatment device can be employed where scale-up of the problem is less. For large-scale attribution, it should be the last resort and can be used as a stopgap arrangement till sustainable alternate solutions are devised.

8.5.1.1 Exploring Possibility of in Situ Treatment of Aquifer

Arsenic contamination is largely along the fluvial tracks in the Ganga-Brahmaputra plains, and its release is due to the result of reductive dissolution of source materials triggered by the microbial activities under the anoxic condition. As long as groundwater contains sufficient dissolved oxygen, arsenic remains fixed in the sediments. These connotations underline the possibility of in situ remedial measure by supply of oxygenated water to the arsenic-contaminated aquifer. In that direction, the Queen's University claim on the development of an eco-friendly in situ treatment of arsenic-contaminated aquifer by injecting oxygenated water into it – which can ensure safe irrigation and potable water supply at an affordable cost – is mentionable.

8.5.1.2 Surface Water as an Alternate Source

Definitely surface water source is the best option instead of exploiting contaminated aquifers if it is available sufficiently both for drinking and irrigation water supply. Surface waters are generally risk-free from arsenic contamination. Few states, namely, West Bengal and Bihar, have adopted some schemes to supply drinking water to the people living in arsenic-affected areas from surface water sources. The question about agricultural water supply still remains.

Rainwater harvesting and groundwater recharge are provoked by many experts as a method for dilution of concentration in the aquifers. Indeed, this innovative approach coupled with rejuvenation of traditional tanks/ponds in the affected areas would not only help in diluting the contaminated water but also to the infiltration of oxygenated water to the contaminated aquifer. A properly conceived 'managed aquifer recharge (MAR)' scheme, which has similar connotation as that of artificial recharge in India with only difference in consideration of water quality and environmental conditions, based on the geomorphology and hydrogeological characteristics of the area, can help reduce concentration level of arsenic in groundwater. MAR is a technique defined as intentional storage and treatment of water in aquifers for subsequent recovery or environmental benefits [9].

8.5.1.3 Possibility of Exploiting Deep Aquifers

In the Gangetic plains, the deep aquifer below 100 m has arsenic safe groundwater, and the deep aquifer is overlain either by clay or impervious formations. CGWB [16] has reported that with proper sealing of the tube well to stop connectivity between the contaminated and non-contaminated aquifer, the deep aquifer can safely be tapped. This approach invokes requirement of aquifer mapping and groundwater modelling study.

8.5.1.4 Bank Filtration (BF) for Sustainable Drinking Water Supply

Riverbank or simply bank filtration (RBF/BF) [18] is a low-cost (pre)treatment technology in which water is withdrawn from wells located close to a river (typically, 20–200 m away). By pumping a bank filtrate well, river water is induced to flow through porous riverbank and bed sediments to the production well. During soil and aquifer passage, chemical, biological and particulate contaminants are removed due to microbial and physical processes and by mixing with groundwater. A schematic diagram of RBF processes is shown in Fig. 8.4.

River water-based supply schemes in India are normally designed for posttreatment of water drawn by intake wells constructed on the riverbed. In such situation, the intake wells only draw water from the river, and considerable amount of money is spent on posttreatment. In the RBF technique, production well draws water both from river and aquifer, and riverbank and bed act as filtering media and thereby reduce drastically the posttreatment cost. The perennial stream/river in and around the arsenic-affected areas can be used for supply of drinking water by employing RBF technique. Riverbank hydraulically connects top aquifer and river. As most of the aquifers up to 20 m depth are free from arsenic contamination, therefore, RBF technique may not have the risk of withdrawing arsenic-contaminated groundwater.

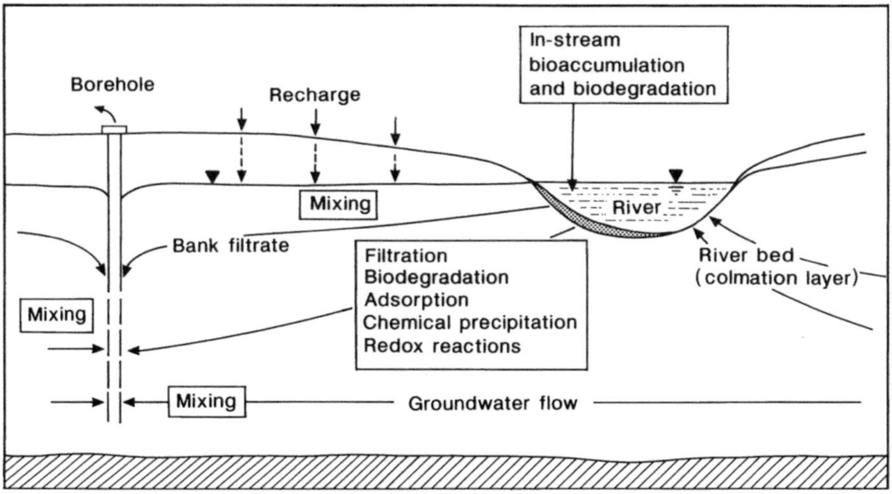

Fig. 8.4 Schematic diagram of processes affecting water quality during bank filtration [11]

Feasibility of implementing a RBF scheme depends on the hydrogeological setup of the area and the physical processes of river-aquifer interaction.

8.5.2 Conservation of Fluoride Contamination

Fluoride problem can be prevented or minimized by using alternate water sources. The preventive measure can be removal of excessive fluoride from drinking water and improvement of the nutritional status of populations at risk. The removal and improvement of the nutritional status can be recognized as living with the problem by care without bothering its mitigation.

It is now well recognized that the source is the fluorine-bearing materials, and their attribution and concentration level depend on the contact time of source material with groundwater. Further, recent aged groundwater (top aquifer) is free from fluoridation, and the risk of fluoridation increases with increasing depth of groundwater because of longer residence time of water with the source material. These connotations underscore the use of surface water, rainwater and low-fluoride groundwater.

8.5.2.1 Surface Water and Rainwater as an Alternate

The use of surface water as drinking water source is the best option in fluoride-contaminated areas. In arid and hyperarid areas, scope may not exist for maintaining perennial condition of surface water sources. Rainwater harvesting can provide a possible alternative. Particular caution is required when opting for surface water, since it is often heavily contaminated with biological and chemical pollutants. Surface water should not be used for drinking without treatment and disinfection.

Rainwater is usually a much cleaner water source and may provide a low-cost simple solution. The problem, however, is limited storage capacity in communities or households. Large storage reservoirs are needed because annual rainfall is extremely uneven in tropical and subtropical regions. Such reservoirs are expensive to build and require large amounts of space.

8.5.2.2 Managed Aquifer Recharge (MAR)

The 'young-aged' groundwater (top aquifer) has less risk of fluoridation because of minimum residence time with the source aquifer materials. An appropriately designed MAR scheme by conserving monsoon surface runoff and groundwater recharge with provision of tapping top aquifer either by the radial collector wells or by open dug wells can help resolve water scarcity problem in fluoride-contaminated aquifer.

8.5.2.3 Bank Filtration (BF) for Sustainable Drinking Water Supply

In areas where perennial stream/river water sources are available, RBF technique can be employed. The same concept as explained in the case of arsenic-contaminated area will apply because top aquifer and river water are risk-free from fluoride contamination.

8.6 Conclusions

Arsenic and fluoride contamination in groundwater are due to the resulting effects of geochemical characteristics of the aquifer materials. The attribution of arsenic in a fluvial track from subsurface sediments containing arsenic-rich iron oxyhydroxides is by the process of reductive dissolution of source materials triggered by the microbial activities under anoxic condition. The attribution of fluoride in groundwater is due to the geochemical characteristics of fluorine-bearing rock-water interaction.

In case of arsenic, oxygen depletion in the groundwater is said to be the cause of activation, and arsenic remains fixed in the sediments as long as groundwater contains sufficient dissolved oxygen. While fluoride concentration in groundwater depends on the reaction time with aquifer materials, i.e. longer residence time of water in the aquifer can lead high fluoride concentration.

Based on the hypotheses given by researchers on origin, occurrence and attributing characteristics of the contaminants in the aquifer being tapped for groundwater withdrawal, possible hydrological and hydrogeological scope for management of contaminated aquifers to help conservation of aquifer from attribution of geogenic contaminants and continuance with withdrawal of groundwater have been suggested; these include:

(i) Preparation of contaminant risk maps including aquifer mapping.
(ii) Adopting innovative techniques for supply of oxygenated water to the arsenic-contaminated aquifer as in situ remediation technique.

(iii) Exploring scope to tap deep aquifer below the arsenic-contaminated aquifer. This can be corroborated by aquifer mapping and groundwater modelling studies.
(iv) Exploring possibility to adopt managed aquifer recharge (MAR) technique by rainwater harvesting with emphasis on tapping top aquifer (young-aged water) as groundwater withdrawal.
(v) In areas where perennial surface water sources, such as river and lake, are available, implementation of riverbank filtration (RBF) or bank filtration (BF) technique can provide a sustainable solution to supply arsenic safe drinking water.

Acknowledgement The author is thankful to the National Academy of Sciences, India, Allahabad, for giving the opportunity to present the paper in its '2nd Policy Support Brainstorming on Safe Water and Environmental Sanitation in Rural and Urban India,' held on 20, 21 and 22 September 2012, at Allahabad, and accepting the paper for publication.

References

1. Acharya SK, Chakraborty P, Lahiri S, Raymahashay BC, Guha S, Bhowmik A (1999) Arsenic poisoning in the Ganges delta. Nature 401:545–546
2. Acharya SK, Lahiri S, Raymahashay BC, Bhowmik A (2000) Arsenic toxicity of groundwater in parts of the Bengal basin in India and Bangladesh: the role of quaternary stratigraphy and Holocene sea-level fluctuation. Environ Geol 39:1127–1137
3. American Dental Association (2005) Fluoridation facts. 69p. http://www.ada.org/~/media/ADA/Member%20Centre/Files/fluoridation-facts.ashx. downloaded on 01/06/2016
4. Bhattacharya P, Chatterjee D, Jacks G (1997) Occurrence of arsenic-contaminated groundwater in alluvial aquifers from delta plains, eastern India: options for safe drinking water supply. Intl Jour Water Res Dev 13(1):79–92
5. Bose P, Sharma A (2002) Role of iron in controlling speciation and mobilization of arsenic in subsurface environment. Water Res 36:4916–4926
6. Brunt R, Vasak L, Griffioen J (2004). Fluoride in groundwater: probability of occurrence of excessive concentration on global scale. Igrac: international groundwater resources assessment Centre. UNESCO. Report No. SP 2004-2
7. Chaturvedi Hema T, Chandrasekharan D, Jalihal AA (2005) Fluoride contamination of groundwater in India – country update. IAH selected papers on Hydrogeology 7. In: Sahuquillo A, Capilla J, Martinez-Cortina L, Sanchez-Vila X (eds) Taylor & Francis
8. CGWB (2011) Dynamic groundwater resources of India. Ministry of Water Resources, Government of India
9. Dillon P, Paul P, Declan P, Helen B, John W (2009) Managed aquifer recharge: an introduction. National Water Commission, Waterlines II, Canberra
10. Guha Mazumder DN (2008) Chronic arsenic toxicity and human health. Indian J Med Res 188:436–447
11. Hiscock KM, Grischek T (2002) Attenuation of groundwater pollution by bank filtration. J Hydrol 266(3–4):139–144
12. McArthur JM, Banerjee DM, Hudson-Edwards KA et al (2004) Natural organic matter in sedimentary basins and its relation to arsenic in anoxic groundwater: the example of West Bengal and its worldwide implications. Appl Geochem 19:1255–1293

13. McAurthor JM, Ravenscroft P, Safiullah S, Thirlwall MF (2001) Arsenic in groundwater: testing pollution mechanisms for sedimentary aquifers in Bangladesh. Water Resour Res 37:109–117
14. Meharg AA, Scringeour C, Hossain SA, Fuller H et al (2006) Code position of organic carbon and arsenic in Bengal Delta aquifers. Environ Sci Technol 40(16):4928–4935
15. Nickson R, McArthur JM, Ravenscroft P, Burgess WG, Ahmed KM (2000) Mechanism of arsenic release to groundwater, Bangladesh and West Bengal. Appl Geochem 15:403–413
16. NIH & CGWB (2010) Mitigation and remedy of groundwater arsenic menace in India-a vision document. Published by National Institute of Hydrology, Roorkee, 184p
17. Ray AK (1999) Chemistry of arsenic and arsenic minerals relevant to contamination of groundwater and soil from subterranean source. Everyman's Sci 35(1)
18. Ray C, Melin G, Linskey LB (eds.) (2003) Riverbank Filtration-Improving source water quality. vol 43. Kluwer Academic Publishers, Dordrecht, 364 p
19. Saha AK (1991) Genesis of arsenic in groundwater in parts of West Bengal. Center for Studies on man and environment, Calcutta, Annual Volume
20. Sexana VK, Sanjeev K, Singh VS (2004) Occurrence, behaviour and speciation of arsenic in groundwater. Curr Sci 86(2):25–30
21. Tushar S (2009) Taming the anarchy: groundwater governance in south Asia. In: Resources for the future. International Water Management Institute, Colombo, Washington, DC
22. Singh Dhirendra K, Singh Anil K (2002) Groundwater situation in India: problems & perspectives. Water Resour Dev 18(4):563–580
23. Smedley PL, Kinniburgh DG (2002) A review of the source, behavior and distribution of arsenic in natural waters. Appl Geochem 17(5):517–568
24. Tanabe S, Saito Vu QL, Hanebuth TJJ et al (2006) Holocene evolution of the Song Hong (Red River) delta system, Northern Vietnam. Sediment Geol 187(1–2):29–61
25. Task Force (2009) Report of the task force on irrigation. Planning Commission. Government of India
26. UNICEF (1999) State of the art report on the extent of fluoride in drinking water and the resulting endemicity in India. Fluorosis and Rural Development Foundation for UNICEF, New Delhi

Narayan C. Ghosh is hydrologist in profession and currently serving as scientist "G" and head of the Ground Water Hydrology Division at the National Institute of Hydrology, Roorkee, India. Dr. N.C. Ghosh graduated in civil engineering from Calcutta University in year 1981 and postgraduated in water resources engineering from the Indian Institute of Technology Kharagpur in year 1984. Dr. Ghosh obtained his PhD degree in civil engineering from the Indian Institute of Technology Roorkee in year 2001. Starting his service career as assistant engineer in the Public Works Department of the Government of Tripura and subsequently working as a research fellow at Jadavpur University, Kolkata, Dr. Ghosh has transformed himself as a specialist in surface and groundwater contaminant hydrology and hydraulics. He had also served as visiting faculty at Tripura Engineering College, Tripura. His research interest and realization mainly focused towards contaminant hydrology and modelling, groundwater flow and transport modelling, stream-aquifer interaction, wastewater reuse and treatment, parameter optimization, etc. He has about 29 years of research/professional experiences and guided 2 PhD theses and several master theses. Dr. Ghosh has published more than 135 scientific papers in international and national journals, conferences, symposia, etc. He has two books and a number of book chapters, scientific reports and project completions, in his credit.

Fluoride and Fluorosis Mitigation: Indian Contributions and Its Impact

9

A.K. Susheela

9.1 Introduction

For the past 25 years, the nation witnessed unprecedented attention and investments for providing safe drinking water both in the rural and urban sectors. The main issue in the drinking water sector is that it should be without biological and harmful chemical contaminants for health reasons.

This presentation is focussing on a chemical contaminant, i.e. fluoride. Fluoride-containing minerals are in abundance in the geological crust of India, paving its way to the underground aquifer. Almost three fourths of a century is over ever since the disease, fluorosis, by drinking fluoride-contaminated water surfaced in India as the first Indian report appeared in 1937 [1].

The presence of fluoride in drinking water and ingestion of fluoride leading to a crippling and debilitating disease, afflicting young and old and men and women of all age groups, are well recorded [2–4].

For almost four decades since the first report on the disease appeared in the *Indian Medical Gazette*, the then prevailing concept was that the negatively charged and highly reactive element, fluoride, causes damage/inflicts injury in calcified/calcium-rich bones and teeth. Therefore, dental and skeletal fluoroses were the known, widely reported disease entities then due to fluoride poisoning [5, 6]. The disease entities attracted considerable attention from the scientific community across the disciplines. The message that the disease has "no treatment or cure" was deeply ingrained in the perception of the medical fraternity. The disease was thus totally neglected. No patient of fluorosis was admitted to a hospital for fear of blocking a bed. This perception has done considerable damage to the victims of the disease as they were the most unwanted and neglected a lot, in the public health domain.

A.K. Susheela (✉)
Fluorosis Foundation of India, B-1 Saransh, 34 I.P. Extension, Delhi 110092, India
e-mail: frnrdf.aks@gmail.com

K.J. Nath, V.P. Sharma (eds.), *Water and Sanitation in the New Millennium*,
DOI 10.1007/978-81-322-3745-7_9

However, during the past three and a half decades, the history has been rewritten due to the monumental contributions that emerged as the result of basic investigations focusing on the matrix molecules of the calcified tissues [7, 8] vs. the non-calcified tissues [9–11] and applied researches on patient samples. The contributions emerged have led to a third entity of the disease, i.e. non-skeletal fluorosis. Detailed scientific information on the non-skeletal, skeletal and dental fluorosis entities of the disease paved the way to develop diagnostic tests for fluorosis, and the disease is diagnosed at very early stages [12, 13]. At times differential diagnosis of fluorosis had to be introduced and practice of interventions by the patients had commenced, culminating in complete recovery in a matter of a fortnight [14, 15].

With this preamble, the chapter is highlighting the major milestones and the landmark events achieved by the nation in fluoride and fluorosis mitigation (Fig. 9.1).

9.2 The Milestones

The result of molecular and biochemical investigations paved the way for reporting a variety of fluoride-caused damages in both soft and calcified tissues of the human body [16].

2.0.	Fluoride destroys soft tissues leading to a variety of health complaints as follows:
2.1.	Fluoride destroys muscle structure and function and depletes muscle energy.
	Fluoride destroys RBCs and adversely affects haemoglobin production [17].
	Fluoride destroys musculature in the wall of blood vessels leading to production of chemical derivatives leading to blocking or calcification of blood vessels.
	Fluoride destroys lining of the stomach and intestine causing gastrointestinal discomfort and non-absorption of nutrients from the diet [18, 19].
	Fluoride impairs carbohydrate and protein metabolism, developing serious abnormalities and functional derangements in the body [20–22].
2.2.	Fluoride toxicity on matrix molecules of the bone and teeth led to significant contributions to negate the belief that fluoride is good for the teeth. Fluoride does more damage to the teeth than any good.
	Dermatan sulphate, a matrix molecule formation in the bone and teeth, does not exist under normal condition. It appears to be the detrimental factor in causing dental and skeletal forms of fluorosis.
2.3.	The effects of fluoride on structural and biochemical entities of soft tissues led to the understanding of the molecular mechanism involved in ectopic/soft tissue calcification known to occur in fluorosis.
2.4.	The revelation of the molecular derangements taken place in fluoride poisoning led to the development of a diagnostic test for fluorosis which comprises of a battery of investigations involving testing of fluoride in (1) drinking water, (2) serum and (3) urine samples besides (4) review of the soft tissue calcification (i.e. interosseous membrane calcification in the forearm) through *X-ray radiograph*. The tests, developed in the early 1990s, have stood the test of time and fine-tuned for almost 20 years. This is the only diagnostic procedure reported to date. In view of the Indian contributions leading to early diagnosis of the disease and practise of inteventions with focus on withdrawl of fluori de from ingestion and promotion of diet rich in nutrients led to complete recovery from the disease. This have had an impact on patients of fluorosis both nationally and globally.

Fig. 9.1 Showing the Foundation's activities are extended to the endemic states of Fluorosis across the nation

The impact of the five major milestones in the national scene has been substantial. India is looked upon by the developed world as the final destination for consultations for the mitigation of fluorosis (Fig. 9.2).

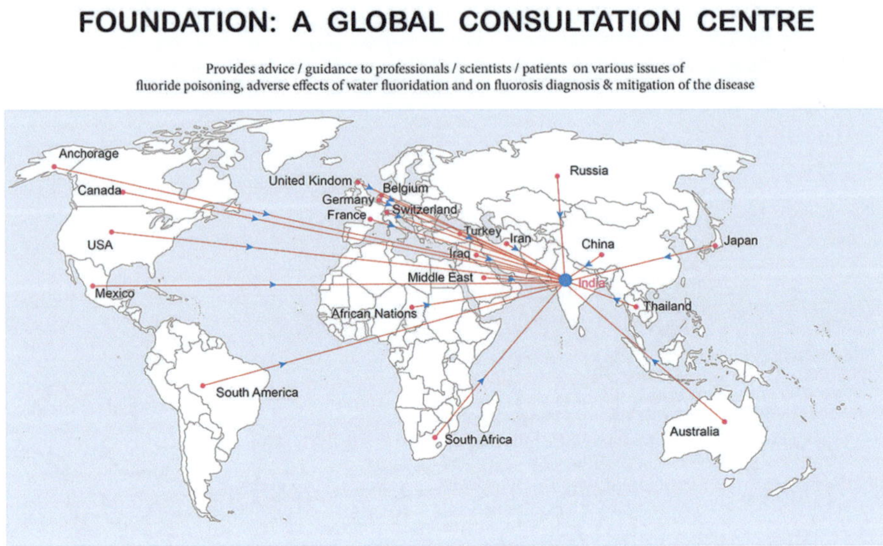

Fig. 9.2 Showing the Foundation located in New Delhi, India is consulted by the patients of Fluorosis, Scientists and Professionals working in field of Fluorosis across the Globe

9.3 The Developments in the Drinking Water Sector in India [23, 24]

(i) All sources of drinking water existing in a village/block/district are to be tested for water quality with focus on fluoride using ion-specific electrode. Identifying safe water (F<1.0 mg/L) sources from fluoride contaminated sources ought to be carried out in all endemic states for fluorosis. All safe sources are to be safeguarded and to be used only for human consumption. To ensure safety, sources should not dry up. They should be recharged through rainwater harvesting by permitting to stagnate the water on the ground and preventing runoff.

(ii) Among the various treatment technologies indigenously developed, (1) Nalgonda and (2) Activated alumina are excellent but not feasible for rural community to carry out operation and maintenance without involvement of a scientific institution for all times and therefore not recommended.

(iii) To introduce reverse osmosis (RO) technology, if not affordable by households, industrial houses have established Community RO plants and provided treated water on subsidised rate to the community.

(iv) The best and long-term solution, however, is to make use of river water (perennial), treat, and supply. Large numbers of state governments are moving in this direction. Excellent examples have already been set up in the country.

9.4 The Developments in Dealing with the Disease in the Health Sector [25]

Information through the print and electronic media on (1) how to suspect fluoride poisoning, (2) how to diagnose and confirm the disease, (3) the interventions to prevent the disease and (4) the procedures recommended to monitor the patient is now available in the public domain. Both the doctors and victims of fluorosis and/or their families are making use of it and reaching out to institutions where they are being cared.

State Health Departments are establishing fluorosis diagnostic facilities in the hospitals. Infrastructure development and manpower training are being taken up, both for conducting tests and providing reliable diagnostic test report to the patient. Such facilities are now available in many endemic states. Some states are in the process of developing such facilities. Investments are being made by State Health Departments.

Directorates of Health Services/Medical Education of the State Health Departments are investing for sensitizing doctors from all disciplines on all aspects of fluorosis, so that the generated awareness changes the unrealistic, unscientific and outdated perceptions of the disease, and victims of fluorosis are truly benefited. From 1993 to 2017, the total number of doctors is sensitized on all aspects of fluorosis in various teaching hospitals, and district hospitals are to the tune of a little over 10000 in endemic states; technical personnel trained for hospital laboratories are 123.

9.5 Emerging New Avenues

Projects centred around fluoride poisoning effects causing damages to maternal and child health resulting in anaemia during pregnancy, low birth weight babies and infant mortality were funded by (1) the Department of Science and Technology (2005–2009) and (2) Indian Council of Medical Research (2011–2013) [26–28].

School health programmes, specially focusing on rectification of anaemia in adolescent girls, have been taken up by the Department of Health Research and Ministry of Health and Family Welfare on high priority. The message that fluoride poisoning is one of the reasons leading to anaemia in adolescent girls, leading to poor performance in school both in studies and extracurricular activities, is also being dealt with. For rectification of anaemia, iron tablets need not be prescribed to children; such shortcuts are not sustainable. The best approach is through nutrition education, with diet editing and diet counselling to mothers of the children.

With such major changes in the understanding of the disease, diagnosis of the disease and prevention and control strategies are being fine-tuned; even victims of fluorosis from overseas are consulting the foundation and following the procedures developed in India. Some overseas patients are seeking appointments to come to India to get the diagnostic tests done and have first-hand information on practice of interventions. These are major developments in the country and the impact is palpable.

9.6 National Fluorosis Control Programme (GOI)

It is also important to point out that since the 11th 5-year plan period (2007–2012), the Ministry of Health and Family Welfare, Government of India, has launched the 13th National Programme for Prevention and Control of Fluorosis. The programme implementers are now beginning to understand the implications and are providing funds to various endemic districts in the country, to plan and launch investigations with adequate infrastructure and manpower to address the issues and reach out to the community with positive messages for control of the disease.

Attention: Clinicians in Government and Private Hospitals [29]

The developments in Health Sciences for early diagnosis of fluorosis and the interventions to practise for complete recovery from the disease have been brought to the public domain. The information announced in print media during the year 2011 is as follows:

- Fluoride poisoning and fluorosis are on the rise in the country; major cause for concern is wrong diagnosis.
- Patients attending OPDs when complaint of the following, ought to be suspected for fluorosis and necessary tests to be carried out.

(1) Irritable bowel syndrome (IBS); (2) polyuria and polydipsia (excessive thirst and tendency to urinate frequently); (3) extreme fatigue/exhaustion/loss of muscle power; (4) insomnia (sleepless nights); (5) low haemoglobin (anaemia); (6) depression; (7) high cholesterol and high blood pressure; (8) joint pain; (9) frequent breaking of bones; (10) disabled children with bowleg, knock-knee, short stature and mental retardation (low IQ), (11) pregnant women with anaemia, low birth weight babies, preterm deliveries, intrauterine death, neonatal death and infant mortality.

It is a fact that fluoride-poisoned patients may not respond to orally administered drugs.

- Without correct diagnosis, if treated with improper drugs, the condition may lead to major complications viz. Renal failure, Blood vessel blockage and Neurological complications.
- The sources of fluoride entry to the body listed below, should be withdrawn forthwith:

Black rock salt containing (1) home-cooked food and/or packaged food products, (2) street food/fruit juices, (3) packaged beverages, (4) spices and (5) fluoridated dental products. Diagnosis of fluoride poisoning/fluorosis at early stages is possible by conducting a few essential tests: The tests are fluoride in (a) serum, (b) urine and (c) drinking water and (d) an X-ray radiograph of the forearm.

- Presently Fluorosis diagnosis is possible in Delhi state at the following Institutions:
  1. All India Institution of Medical Sciences, New Delhi – 1100 29
  2. Fluorosis Foundation of India, New Delhi – 1100 92
  3. Maharishi Valmiki Hospital, Pooth Kurd, Delhi – 110062
- The disease has no treatment; recovery is through practising two interventions which include diet editing and counselling with proper nutrition education.
- This is to announce the opening of the *Rhyon International Fluorosis Diagnostic and Counselling Centre* under the Fluorosis Foundation of India. The services are open to the patients from India and around the globe.

It is open from Monday to Friday from 9:30 to 6:00 pm, except govt. holidays.

9.7 Concluding Remarks

One of the most important messages that need to be highlighted is that the programme on fluorosis mitigation has developed to the present stage without major international funding except for funds from International Development Research Centre (IDRC) Ottawa, Canada, for a project "Fluorosis–India" where three institutions participated and was coordinated by the author of this publication then at AIIMS, New Delhi, during 1980–1983. This laid the foundation for all the developments that took place in India during the later years. UNICEF India and WHO India Country Office also supported the programme. Developments taken place during subsequent years are through funds provided by the National Exchequer. Paucity of funds may have delayed the developments, but the rewarding experience is that the final goal has been achieved to reach out to the community/victims of the disease. Patients of fluorosis are scientifically and accurately diagnosed and addressed to their benefit and advantage.

Acknowledgements The author wishes to acknowledge the support received from IDRC, Canada, and various National Funding Agencies for the projects implemented for a period of over four decades (1970–2017). A large number of scientists, research associates and graduate students worked on the subject "fluoride and fluoride toxicity" and contributed substantially to the understanding of the disease. They are gratefully acknowledged. The experts both from the health and drinking water sectors in the country that worked closely with the foundation are being thanked for their contributions. Lastly all those patients, who provided samples for investigations for decades and led the way for the emerged scientific contributions and developments taken place in India, reported in this communication, are remembered and thanked.

References

1. Shortt HE, Pandit CG, Raghavachari TNS (1937) Endemic fluorosis in the Nellore district of south India. Ind Med Gaz 72:396–398
2. Singh A, Jolly SS, Bansal BC (1961) Skeletal fluorosis and its neurological complications. Lancet 1:197
3. Singh A, Jolly SS, Bansal BC, Mathur OC (1963) Endemic fluorosis. Medicine 42:229
4. Jolly SS, Singh ID, Prasad S, Shama R, Singh BM, Mathur OC (1969) An epidemiological study of endemic fluorosis in Punjab. Indian J Med Res 57:1333
5. Joly SS, Singh BM, Mathur OC, Malhotra KC (1968) Epidemiological, clinical and biochemical study of endemic dental and skeletal fluorosis in Punjab. Br Med J 4:427
6. Krishnamachari KAVR, Krishnaswamy K (1973) Genuvalgum and osteoporosis in an area of endemic Fluorosis. Lancet 2:877–879
7. Susheela AK, Sharma K, Rajan BP, Gnanasundaram N (1988) Human Dental Fluorosis: the status of sulphated isomers of Glycosaminoglycans. Arch Oral Biol 33(10):765–767
8. Susheela AK, Bhatnagar M, Gananasundaram N, Saraswathy TR (1999) Structural aberrations in human fluorosed teeth: biochemical and scanning electron microscopic studies. Curr Sci 77(10):1677–1681
9. Kharb P, Susheela AK (1994) Fluoride ingestion in excess and its effects on organic and certain inorganic constituents of soft tissues. Med Sci Res 22:43–44
10. Susheela AK, Kharb P (1990) Aortic calcification in chronic fluoride poisoning: biochemical and electron microscopic evidence. Exp Mol Pathol 53:72–80
11. Susheela AK, Bhatnagar M, Vig K, Mondal NK (2005) Excess fluoride ingestion and thyroid hormone derangements in children living in Delhi, India. Fluoride 38(2):98–108
12. Susheela AK, Das TK, Khurana JS, Jayaswal A, Dave PK (1988) Circulating levels of sialic acid and glycosaminoglycans: a diagnostic test for ankylosing spondylitis. Ann Rheum Dis 47:833–837
13. Susheela AK (2005) Fluorosis: Doctor's handbook, Ministry of Health and Family Welfare (GOI) sponsored and Printed by the WHO India Country Office, New Delhi
14. Susheela AK (2001) Fluorosis – early detection and management: importance of practising interventions with focus on consuming safe water and essential nutrients. Heinz Nutritional Foundation India, In-Touch 3(3):2001
15. Susheela AK, Bhatnagar M (2002) Reversal of fluoride induced cell injury through elimination of fluoride and consumption of diet rich in essential nutrients and antioxidants. Mol Cell Biochem 234/235:335–340
16. Susheela AK (2007) Treatise on fluorosis: Rev. 3rd edn. Fluorosis Research and Rural Development Foundation, Delhi
17. Susheela AK, Jain SK (1986) Fluoride toxicity: erythrocyte membrane abnormality and "echinocyte" formation. Fluoride Res Stud Environ Sci 27(99):231–239
18. Dasarathy S, Das TK, Gupta IP, Susheela AK, Tandon RK (1996) Gastro-duodenal manifestations in patients with skeletal fluorosis. J Gastroenterol 31:333–337
19. Das TK, Susheela AK, Gupta IP, Dasarathy S, Tandon RK (1994) Toxic effects of chronic fluoride ingestion on the upper gastrointestinal tract. J Clin Gastroenterol 18(3):194–199
20. Jha M, Susheela AK, Krishna N, Rajyalakshmi K, Venkiah K (1982) Excessive ingestion of fluoride and the significance of sialic acid upon glycosaminoglycans in the serum of rabbit and human subjects. J Toxicol Clin Toxicol 19(10):1023–1030
21. Susheela AK, Jha M (1982) Fluoride ingestion and its influence on Glycosaminoglycans in cancellous and cortical bone – a structural and biochemical study. Fluoride 15(4):191–198
22. Susheela AK, Sharma YD (1982) Certain facets of Fluoride action on collagen protein osseous and non-osseous tissues. Fluoride 15(4):177–190
23. Handbook on Planning and Implementation of Water Supply Programme in Fluorosis Endemic Areas in India. A Multi-Author Contribution. In: Susheela AK (ed)., Published by Fluorosis Research & Rural Development Foundation, Delhi – 92, 2009

24. Susheela AK (2011) Fluoride and fluorosis mitigation in rural India: why state governments are failing in their efforts and investments to achieve the goals: the way forward. In: Proceedings of the National workshop-cum-exhibition on drinking water quality. Published by Ministry of Drinking Water and Sanitation, Government of India, New Delhi, 2011
25. Susheela AK (2002) Fluorosis in developing countries: remedial measures and approaches. Proc Indian Natl Sci Acad (PINSA) B68:389–400
26. Susheela AK, Mondal NK, Gupta R, Ganesh K, Brahmankar S, Basin S, Gupta G (2010) Effective interventional approach to control anaemia in pregnant women. Curr Sci 98(10):1320–1330
27. Susheela AK, Mondal NK, Gupta R, Ganesh K, Brahmankar S, Basin S, Saxena A (2010) A novel and effective interventional approach for prevention and control of anaemia in pregnancy and low birth weight babies. In: Global Maternal Health Conference Organized by Bill and Melinda Gates Foundation; Engender USA and Public Health Foundation of India, New Delhi
28. Susheela AK (2010) Anaemia in Pregnancy: an easily rectifiable problem. Guest editorial. Fluoride 43(2):104–107
29. Fluoride Poisoning and Fluorosis (2011) Attention clinicians in government and private hospitals. Times of India, July 2011 (Country-wide announcement)

Dr. A.K. Susheela has worked with the most distinguished medical professionals in the developed world during her postdoctoral assignments in the UK, the USA and Canada. She returned to India in 1970 to join the faculty of the prestigious All India Institute of Medical Sciences, New Delhi. After a glorious, brilliant academic career spanning over a period of three decades and upon superannuation, she took up the reins of the Fluorosis Foundation of India, as its first executive director.

She has raised the foundation to great heights through her orbit-shifting innovations in dealing with health issues and welfare of the community. The foundation is a national and global consultation centre. She is an authority on fluorosis and associated health issues. Her innovative and path-breaking discoveries addressing correction of anaemia in pregnancy, improved birth weight of newborn and reduction of disabilities in infants and children have attracted attention from the world over. She is not only a scientist of high calibre and repute but passionately involved in her scientific pursuits. She is an institution by herself.

Community-Based Approach for Mitigation of Arsenic Problems: Case Studies in West Bengal, India

10

Anirban Gupta, Sudipta Sarkar, Debabrata Ghosh, and Kabita Maiti

10.1 Introduction

The dwindling availability of safe drinking water is the cause for concern for the international scientific community. Right to safe water has been reiterated in the Millennium Development Goal (MDG) of the United Nations stressing the issue as fundamental right to the people. Water quality crisis is acute in many countries including India. Since the last few decades, arsenic contamination in drinking water of millions of people living in different parts of the world has been in the focus of attention of public health scientists and engineers. Several states of India are affected by the toxicity posed by arsenic in the groundwater used for drinking and agricultural purposes. The arsenic contamination in drinking water, fueled by prevailing malnutrition in general rural people, resulted into the manifestation of several diseases, many of which are fatal. It caused a serious predicament to the lives and livelihoods of the local water users since these diseases and arsenic contamination create a cause of social, economic, and ecological imbalance [1–4].

The use of groundwater in these regions is favored by its easy availability and microbial safety. While the best solution to the crisis is to switch over to treated surface water where there is no arsenic contamination, development and maintenance of

A. Gupta (✉) • D. Ghosh
Civil Engineering Department, Indian Institute of Engineering Science and Technology (IIEST), Shibpur, Howrah, West Bengal 711103, India
e-mail: anirban@civil.iiests.ac.in; ghoshdebabrata46@gmail.com

S. Sarkar
Department of Civil Engineering, Indian Institute of Technology (IIT) Roorkee, Roorkee, Uttarakhand 247667, India
e-mail: srkarfce@iitr.ernet.in

K. Maiti
Shibpur Association for Technological, Humanitarian and Environmental Endeavours (SATHEE), Howrah, West Bengal 711103, India
e-mail: m.kabita6@gmail.com

K.J. Nath, V.P. Sharma (eds.), *Water and Sanitation in the New Millennium*,
DOI 10.1007/978-81-322-3745-7_10

surface water-based drinking water system are expensive, time-consuming, and investment intensive. With all these difficulties, it is unlikely for a developing country like India to switch the source of drinking water from groundwater to surface water within a short period of time. In order to save lives before such a switchover is possible, it is imperative to build sustainable arsenic removal systems in an urgent basis. During the last two decades, several technologies for selective removal of arsenic from the contaminated water have been developed, and many of the technologies have matured to be commercially available in the field [5–25].

In view of the dire consequences of fatalities related to arsenic contamination and understanding the public health emergency, the Government of India formulated intervention policy to bring in proper mitigation and prevention measures to resolve the water quality crisis. Following a supply-driven approach, the government started installing community-based arsenic removal units in the worse affected areas with the help of its departments and technical agencies; in many instances, domestic (point-of-use) arsenic filters were also supplied to the individual families. However, the model was not participatory; the government paid for the cost of the treatment units. However, the supply-driven initiative could not motivate the community's involvement in the program. And, eventually in most of the cases, the benevolent actions of the government failed to make any dent on the problems being faced by the affected villagers [26].

Contextually, it is noteworthy; National Water Policy [27] laid emphasis on community-based approach, need-based support, and maintenance of water quality, which all have been delineated as a preemptive need. According to the policy, "the local initiatives requires to strengthen 'Right to water' issue, to rejuvenate the traditional techniques in conservation and management, to recognize water as a common property and not merely a commodity." It is perceived that people-centered and community-based water management should get the highest priority as it contributes to the national goals of the promotion of safe water arrangement [27].

Considering quintessence of the National Water Policy, area-specific crucial predicament and argumentation over downsides of supply-driven initiative, IIEST, Shibpur (erstwhile, Bengal Engineering and Science University (BESU), Shibpur), conceived and implemented a project in which there were two crucial components: first, to develop and implement an innovative, user-friendly, and cost-effective solution for the promotion of arsenic-safe water arrangement for large social groups (at the community level) in rural and peri-urban settings and, second, to develop a social institution at every location which shall be crucial for making a strategic shift from the supply-driven initiative to a demand-driven self-help model. This article provides an account of the successful technological interventions made in various communities in West Bengal, India. This chapter also describes the details of the community-centered participatory model as an approach to manage the operation and maintenance of the treatment units. The authors also discuss the effects of such socio-technological intervention in achieving benefits that are far beyond merely the remediation of arsenic problem.

10.2 Arsenic Remediation: Choice of Scale of Operation

Success in field-level implementation of a treatment process developed in a laboratory depends on many factors. One such factor is the choice of appropriate scale to which the technology is to be upgraded to. For field-level implementation in the rural areas in developing countries, there may be two different types of applications. One of them is the point-of-use (PoU)-type household treatment unit, and the other is a community-scale arsenic removal device that can cater to a whole community or a group of neighboring villages. The household water treatment units have modular design, portability, and ease of installation and use. They offer flexibility of operation and maintenance that can be dictated by the owner himself. One community-based water treatment unit, based on its size, can replace installation of hundreds of individual household arsenic removal units. Although the household treatment units offer several advantages over a community-based water treatment unit, for arsenic removal purposes, the latter is chosen due to their following advantages:

- Treated water from any arsenic treatment unit should be tested for arsenic at predetermined time intervals to ensure reliability in operation and quality control. Frequent determination of arsenic in treated water is cost prohibitive for an individual using a household treatment unit. For every single analysis of a water sample from a community-based unit, hundreds of water analyses would be required for individual household units, all other conditions remaining identical.
- Household PoU units using adsorbent media or coagulants will always generate arsenic-laden sludge or solids. Coordinating collection and safe disposal of sludge or used media from individual households poses a level of complexity and enforcement effort that are difficult to sustain in remote villages. The well-head community-based units reduce such management problems by manyfold.
- For an arsenic removal unit using adsorbent media, replacement or regeneration of exhausted adsorbent can be cost prohibitive for an individual using household. But for a community-scale unit, collective effort from the users makes it affordable to replace or regenerate the exhausted adsorbent media.
- Even for household units, the villagers (mostly village women) need to come every day to the existing wells to collect water. Thus, having a simple-to-operate arsenic removal system mounted on the well provides continued collective vigilance with regard to color (caused by iron), smell (due to biological activity), or other abnormal behaviors that may otherwise go unnoticed for household units.
- Cost is one of the key parameters involved in the choice and adoption of a technology by a community. Both fixed and operating costs of the community based well-head unit are significantly lower than the total sum of the same for the household units that it replaces. Also, it is relatively easy to introduce modifications and innovations in well-head units for performance enhancement through the collective and direct involvement of the community through a villagers' appointed water committee.

Considering the above arguments, we decided that community-based arsenic removal units are better suited for the purpose than the individual household units.

10.3 The Treatment Unit and Its Performance

In order to qualify as an appropriate technology in a developing country, while designing the treatment units, one needs to provide special considerations for conserving the traditional practices and values while ensuring that the local needs are effectively satisfied. Figure 10.1a is a photograph showing the usual method of collection of water. Figure 10.1b shows a photograph of the treatment unit in use by a villager. It may be noted that installation of the arsenic removal device did not change the villagers' water collection practices; the only difference is that they now collect water at the exit of the treatment unit instead of collecting water from the spout of the hand pump, as practiced earlier. Thus, the technological solution did not call for a change in the traditional ways of collection of drinking water, and it delivers arsenic-safe drinking water. As a result, there was an immediate acceptance from the users. The treatment unit consists of a gravity-flow adsorption column which housed 100 kg of activated alumina as adsorbent. The unit was adequately designed to support the regular operation as well as the maintenance operations such as backwashing operation and adsorbent media replacement after the adsorbent gets exhausted once in a year. The detail design of the units can be found elsewhere [28–30]. The operation of the unit is simple and user-friendly. The flow rate of the treated water is 10 l per minute and has been satisfactory for the villagers. Apart from arsenic removal, unique

(A) (B)

Fig. 10.1 (**a**) Usual method of water collection in the villages of West Bengal, India (**b**) The method of collection of water after the treatment units is installed

design of the treatment unit allows the atmospheric air to enter the unit during its operation, which helps in oxidizing the ferrous iron present in water to form insoluble ferric hydroxide particles. The ferric hydroxide particles are removed by straining action of the adsorbent bed. Thus, the arsenic removal units achieve a high degree of removal of dissolved iron present in the raw water improving thereby the palatability and general acceptance of the water. Figure 10.2a, b show the performance of one of the treatment units for removal of arsenic and iron, respectively.

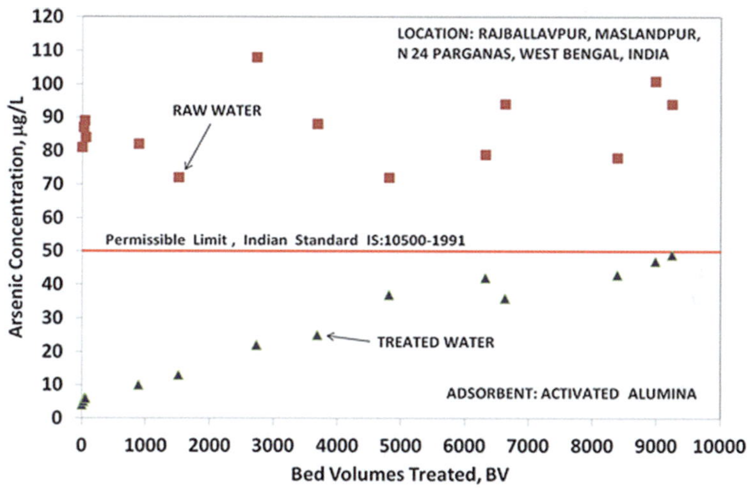

(A)

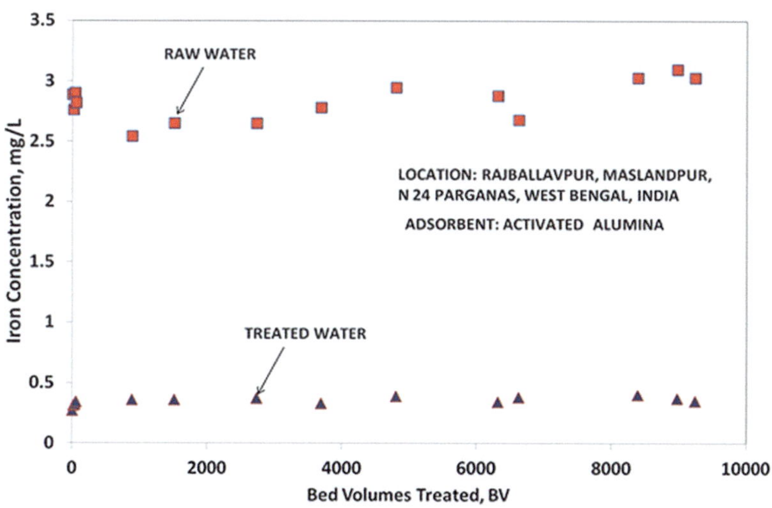

(B)

Fig. 10.2 The performance of the arsenic removal unit located at Rajballavpur village in Maslandpur, North 24 Parganas in West Bengal; influent and effluent histories showing (**a**) arsenic and (**b**) iron removal

The treatment unit is fabricated locally with locally available materials including the adsorbent (activated alumina). The units are robustly built and do not require regular maintenance. The adsorption column acts as a plug flow reactor. Therefore, it is forgiving toward any occasional increase in arsenic concentration in the raw water. Unlike batch-type reactors which are mostly used for coprecipitation method of arsenic removal, these units consistently produce arsenic-safe water in a reliable manner. Reliability of operation and consistency in producing arsenic-safe drinking water are necessary requirements particularly in rural settings in a developing country where frequent measurement and detection of arsenic in treated water is difficult and cost prohibitive.

Upon exhaustion of the adsorption column, media from each unit is replaced by previously regenerated media. The exhausted media is taken to a central regeneration facility where it is regenerated with adequate precaution and arrangement to contain any arsenical waste. The beneficiaries (concerned water committees) provide the necessary cost and charges of regeneration.

10.4 Regenerable Versus Once-Use Adsorbent

Economy in operation is the major reason behind choosing an adsorbent that can be regenerated and reused over many cycles. This is especially true in the developing countries where, in order for a technology to be acceptable to the masses, it has to be affordable. A significant amount of savings in terms of cost of treated water is possible when a regenerable adsorbent is used instead of onetime-use or throwaway adsorbent. The savings are more in the developing countries where labor and chemicals, the two major resources required for regeneration, come at a lower price compared to the developed world. Moreover, the regeneration procedure being simple and easy, even a group of trained villagers with little or no prior technical knowledge can effectively carry out the whole regeneration process. The cost of regeneration is just one-fourth of the price of fresh activated alumina. After the regeneration, high concentration of arsenic in the spent regenerants is separated from the liquid phase forming a solid residue with ferric hydroxide particles. The resultant spent regenerants are innocuous; it does not contain arsenic at concentration higher than the regulatory limit set for the wastewater.

One of the important parameters for sustainable arsenic remediation is the proper management of treatment residuals. The treatment residual generated in this process is rich in arsenic. Hence, they need suitable storage and disposal so that any leaching of arsenic can be avoided. Had the adsorbent been a onetime-use or throwaway adsorbent, the amount of the treatment residual generated would have been about 100 L in volume or about 100 kg in mass. However, the treatment residual generated in the process of regeneration followed by the arsenic sludge formation, as discussed above, does not have mass not more than 2 kg. Therefore, it is easy to handle and manage this small amount of treatment residual as opposed to 100 kg of treatment residual in case of non-regenerable adsorbent. The cost of disposal of treatment residuals, if taken into account in the calculation of overall cost of project and the treated water, the onetime-use adsorbent will be a worse choice. Moreover, in the countries where the hazardous waste regulations are not enforced properly, the possible mismanagement that may arise due to handling of large amount of treatment residuals would cause a menace that is comparable with the arsenic crisis in the groundwater itself, if not worse. Thus, the use of regenerable adsorbent was a better choice for the project. The authors also developed simple, easy-to-be-followed technology for safe storage of the solid treatment residual or the sludge so that under normal environmental conditions, there is no significant leaching of arsenic. Unless this particular waste management part is not taken care of, the highly toxic arsenic species that was removed in the treatment process would get mobilized in the environment again, thereby defeating the whole definition of sustainability.

10.5 Community-Based Management: Comparison Between Approaches

The government-initiated community-based remediation effort is a supply-driven approach. Understanding the public health emergency, the community and domestic filters are being provided to the people in the affected areas for free of cost. The government-initiated programs did not yield expected results, and many of the treatment units are now lying defunct due to the reason that the community did not come forward to take care of the treatment units. Thus, there are enough examples to doubt on whether a supply-driven approach taken by the government could influence over the community's involvement in the arsenic remediation program. The supply-driven mode in the government program is reminiscent of establishing a relation between the giver and taker; it does not make the people adequately aware about the values associated with such benevolent action. It does not either enrich the beneficiaries with the feeling of self-reliance. On the contrary, the long-term sustenance of the program demands that the community should grow up with awareness and knowledge, participate in the operation, maintenance and upkeep of the installed units, and finally bring in innovative solutions to make things better.

The above logical argumentation does not necessarily conclude that a demand-driven approach, instead, may be successful. Arsenic in water does not cause any physical changes in the esthetic properties of water, such as color, taste, temperature, appearance, etc. This is a public health crisis, and many of the affected communities do not have the right kind of awareness about what causes the severe health crisis and what are the technological options they have for remediation of the problem. Unless such awareness is generated, the demand in the community for having an arsenic removal unit does not evolve. Also, people are not in the right economic strata to fund the capital cost of such devices. Thus, a fully demand-driven approach also shall be far from solving the water quality crisis. Anticipating the problems of imbalances between supply- and demand-driven mode and realizing the potentials of a balanced approach, IIEST adopted a hybrid mechanism where the initial capital investment comes from outside the community and the beneficiaries within the community make consistent logical contributions from which the cost of operation and maintenance can be arranged. This self-help approach is introduced in IIEST's program. The approach can motivate the users to have collective ownership that brings in pride and commitment which ultimately stimulates effective participation in proper use and maintenance. As an outcome of this social process, the ordinary beneficiaries and even the passive receivers both become effective facilitators as well as rational recipients.

10.6 IIEST's Intervention in Safe Water Management

It has been realized that the decentralization in water management system and the empowerment of the local users can strengthen the survival base of the rural economy. Rural water supply should not be treated as a mere service delivery process but as a step toward promoting water security in a community. Active participation of a community is required in every aspects of the rural water supply system, right from the planning to the implementation stage and also during the regular operation and maintenance. As such, community participation has long been recognized as an effective means of helping rural and urban people to focus energy and mobilize resources to solve their health, environmental, and economic problems.

Before it was even decided to install a treatment unit in a village, authors initiated the process of awareness generation within the community about the arsenic-related health problems, its origin and effects, ultimate fate, and the importance of arsenic-safe water to solve the health problems. Such efforts helped us to create awareness for drinking arsenic-safe water. The community also was informed about the role of arsenic remediation units to produce arsenic-safe drinking water. The information that there exist effective ways to treat arsenic-contaminated water generated demands for having arsenic removal unit in their community. The target communities are also informed of the proposed technology, the management system, their duties, collective ownership, need for paying the water tariff, short- and long-term benefits of arsenic treatment systems, and also about the importance of their active participation for sustaining the operation and maintenance. Such informal education was sensitized through the use of group meetings by the activists, paper pamphlets, audiovisual communication including movies and street-corner short plays, etc. Figure 10.3 shows a photograph of a street-corner short play being held in one of the affected villages.

In order to ensure community participation, we involved the community right from the planning stage leading to the installation and maintenance of the treatment units. Through participatory rural appraisals, the villagers make decisions whether and where to install the arsenic removal units. The villagers form a water committee comprising of local residents of which at least one-third members are women. After installation and initial monitoring, the ownership of the unit is transferred to the villagers. The duties of the members of the water committee are distributed, and they meet at least once in 2 months to discuss about the issues being faced. Usually, each committee appoints one or two local youths as caretakers who take part in daily backwashing, operation, and upkeep of the unit. Suitable operation and

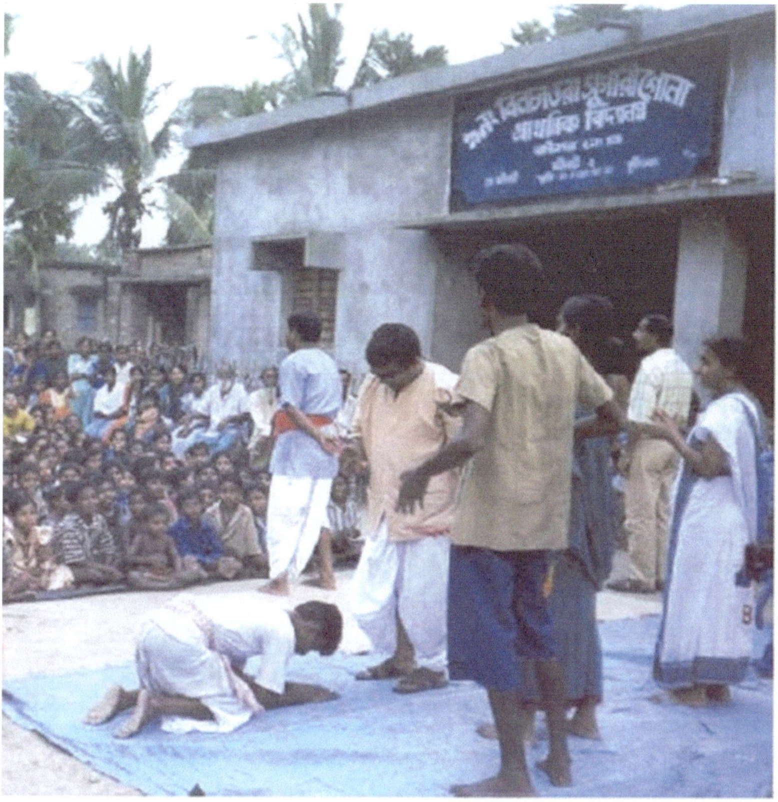

Fig. 10.3 Photograph of a street-corner short play being enacted in the villages for awareness generation

maintenance manuals have been given to all the water committees, and the members of the water committees and the caretaker have been trained for upkeep/maintenance of the filter and maintenance of accounts for the collected water tariff. The water committee collects monthly water tariff from the user families, updates the tariff cards, pays for the salaries of the caretakers, pays for monthly water test, and holds the money in a nearby bank for future expenditures. The members of the water committees are nominated/elected every year from among the users. In case of any emergency related to water quality, apart from contacting the local plumbers who are trained to undertake mechanical maintenance, the water committees could directly contact local representatives of the project implementation team.

Though the capital cost of installation of the treatment units was sourced from funding agencies; however, the subsequent costs of operation were not supported by them. Thus, the expenditures related to the operation and maintenance of the unit, including the salary of caretaker, once-a-month water analysis, incidental maintenance, and once-a-year regeneration of the unit, are to be met by the villagers who get benefitted from the treated water. In order to attain long-term economic sustainability for the units, it was necessary to meet the cost of treated water by collection of water tariff, a concept never practiced before in the villages of West Bengal and perhaps elsewhere in India. The water tariff being provided per family per month is only Rs. 15–25. The amount is low when compared to the amount that people usually spend for entertainment and even for cell phones. Still, there was an initial resistance from the villagers to pay the water tariff. It was not easy to introduce this new concept of water tariff in the villages where people are used to get drinking water for free. However, through persistent communications, the villagers were motivated to understand that the amount of water tariff is negligible when it is compared to the possible gain from the recovery of workdays lost on the account of illness. Once started, the concept became popular within a short period of time. The villagers understood that owning and maintaining a community water treatment unit through collective payment in the form of water tariff was a good investment for the health and prosperity of the present and future generations. The return on the investment in the form of savings made on the cost of medicine was immediately realized when the users switched over from contaminated water to the treated water. Obviously, the realization about usefulness of paying for one's own drinking water came after intense awareness generation efforts were undertaken by the project team.

10.7 Toward Sustainability

In 1996 IIEST joined hand with an international organization "Water For People" to develop arsenic removal units, which was appropriately responsive to the dire situation of many parts of West Bengal. Initially, the installation of units focused on sporadic approach; total village solution was not there but some scattered fixes were in place; but in later stage, IIEST transformed the sporadic approach to an integrated approach, wherein a larger section of the community was accommodated in safe water drive to end the water poverty cycle.

Since inception, 209 arsenic removal units, of which 144 units at community level and 65 units at schools in 3 districts (North 24 Parganas, Nadia, Murshidabad) of West Bengal, have been installed. The first community-level unit was set up in 1997, while school-level unit was first installed in 2007, and the installation of new ones is still continuing, which aggregated to a total investment of about Rs. two crores. More than two lakh people are getting benefits out of the safe water drive, and millions need to be reached out. Shibpur Association for Technological Humanitarian and Environmental Endeavours (SATHEE) is partnering with IIEST to reinforce the community management and sustainability of the installed units since 2007.

Detailed data from about 75 such arsenic removal units in different villages/towns in the districts of North 24 Parganas and Nadia of West Bengal are available, and these are presented here. The oldest existing unit is supplying safe water for more than 17 years. Several such units have successfully demonstrated sustainable operation and maintenance effort by the communities and are working incessantly for significant number of years. Figure 10.4 shows a chart showing the time periods through which the units are in operation in these two districts. The number of user families attached with each arsenic removal unit varies depending on several local factors such as affordability, motivation level of the users, prevalence of arsenic-related ailments, etc. Figure 10.5 shows the distribution of the units according to the number of user families attached with them. In many places, the total fund generated from the collection of water tariff far exceeds the annual expenditure associated with the operation and maintenance of the units. The savings thus generated are kept in a bank account maintained by the village water committees. Figure 10.6 shows a chart showing distribution of the units according to the annual revenues earned by the respective water committees. Figure 10.6 is a proof of the economic self-reliance gained by the treatment units, which is a reflection of long-term economic

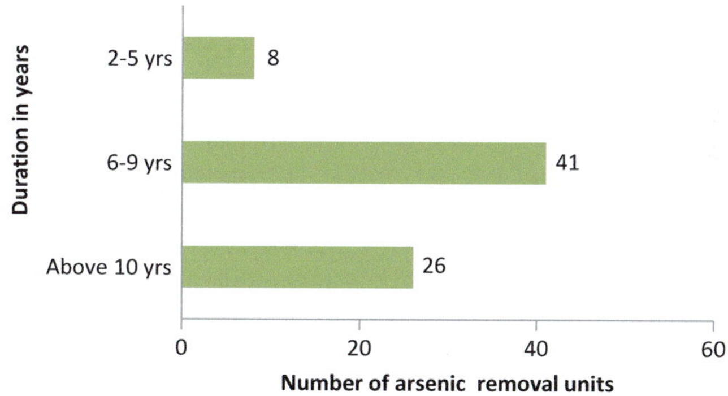

Fig. 10.4 Distribution of the number of years of operation of the arsenic removal units (data from 75 units in North 24 Parganas and Nadia)

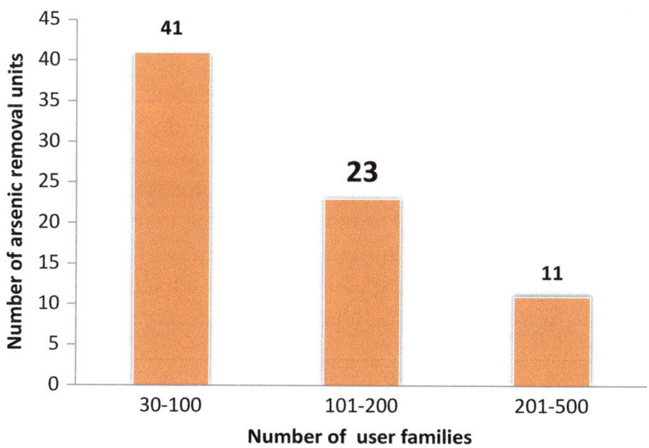

Fig. 10.5 Distribution of the number of arsenic removal units with respect to the number of user families associated with each of them

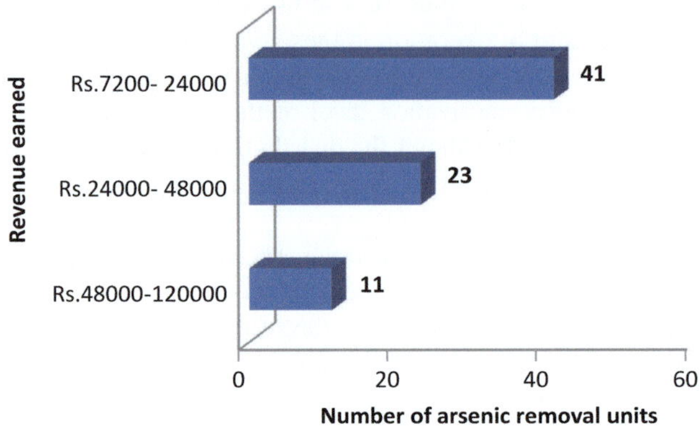

Fig. 10.6 Distribution of revenue earned by each unit with respect to the number of arsenic removal units

sustainability. Once the community becomes confident on managing the water supply system through the participatory model, they venture to take the experience for the community development beyond merely solving the arsenic crisis. In many places, the users decided to upgrade the hand pump-based manual unit to a motor-operated unit so that the collection process becomes easier for them. Not only this, the water committees built overhead storage tanks and provided enclosure and shed over the units so as to facilitate all-weather operations. Figure 10.7 shows the photographs of an arsenic removal unit located at Nabarun Sangha, in Ashoknagar in West Bengal; the two photographs as in part (a) and (b), taken in a gap of 4 years, demonstrate the changes that took place over time. Figure 10.7a shows the bare treatment unit and concrete platform provided in the beginning of the project in 2005. Figure 10.7b shows that there were overhead storage tank, enclosures, all-weather shed, and an electrical pumping system, all installed by 2009. All these additional facilities were made possible with the help of the revenue generated from the water tariff that came from the community itself. In turn, the number of user families also increases and so does the revenue generated. Figure 10.8 shows how the number of user families gradually increased over the period of time since its installation in 2004. Figure 10.9 shows the continuous increase in the revenue earned over this time period. As per our information, among the 75 units for which regular data are being collected, at least at 20 treatment units, the water committees have upgraded the facilities at their units by spending Rs. 30,000–80,000 from their accumulated fund. All these are fantastic examples of what a successful community participation can achieve for a community-level development project.

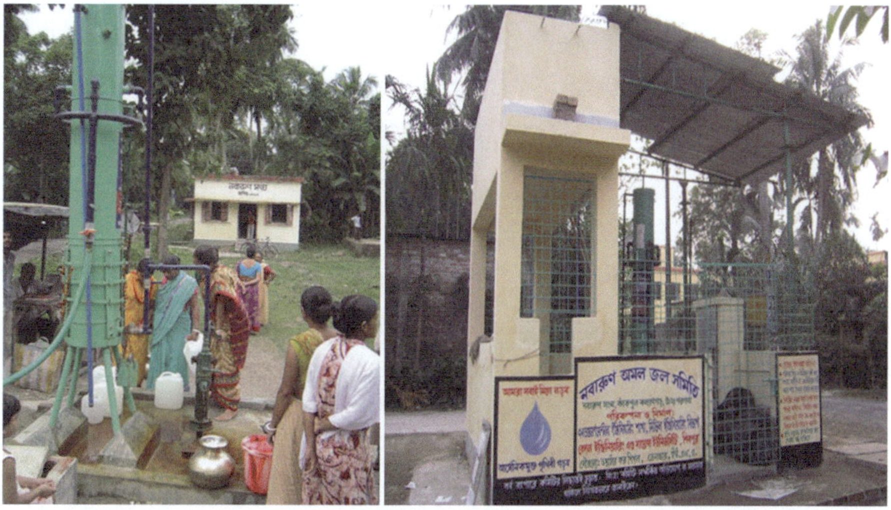

(a) (b)

Fig. 10.7 (**a**) The photograph of the arsenic removal unit at Nabarun Sangha, North 24 Parganas, in 2005 shortly after installation and (**b**) photograph of the same unit in 2009 following continuous upgradation by the village water committee

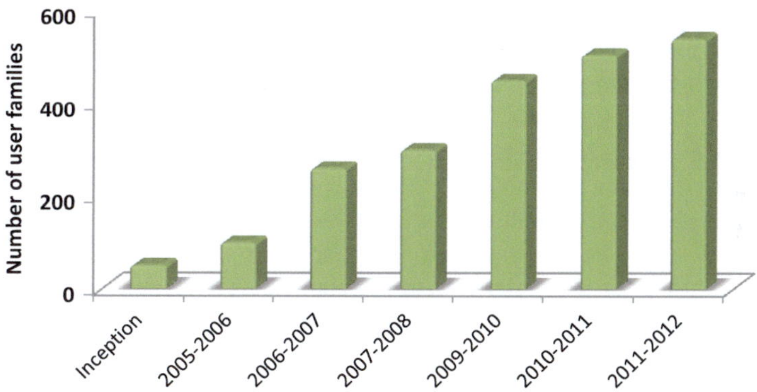

Fig. 10.8 Gradual increase in the number of user families over the last 7 years for the arsenic removal unit located at Nabarun Sangha, Ashoknagar, North 24 Parganas in West Bengal

10.8 Beyond Sustainability

The treatment unit and the social institution for managing the upkeep of the units, both are instrumental for the success of the arsenic remediation program. The reason for overwhelming community participation and support was the short-term as well as long-term benefits brought to the society by the arsenic mitigation program.

The short-term benefits are well identified. The arsenic-related health hazards, before becoming fatal, affect the victim through various manifestations. The victims

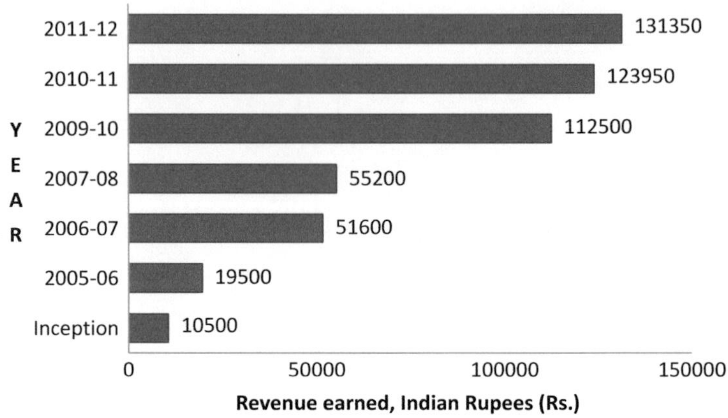

Fig. 10.9 Gradual increase in the revenue earned at the arsenic removal unit in Nabarun Sangha at Ashoknagar, North 24 Parganas

lose ability to work and earn their livelihood and have to buy medicines for cure. However, the first step to stop arsenic-related health hazards starts with switching to arsenic-safe drinking water. Thus, provision of arsenic-safe drinking water not only saves lives directly, but it also saves families from being trodden down through the spirals of poverty. Provision of safe drinking water helped the community to recover the lost workdays on the account of illness. Thus, the provision of a well-managed water system helps the community to generate wealth to prosper. These are measurable benefits to the individual water users.

Moreover, the socioeconomic institution formed for the upkeep, operation, and maintenance of the arsenic removal units provides direct and indirect employments. The jobs are created in every step of the system starting from the fabrication of the arsenic removal units to manufacture of adsorbents. Installation of every unit in the village requires at least 10 labor workdays. Each unit employs one or two persons, on part-time basis, for its operation and maintenance. Also, a group of four trained villagers work as entrepreneurs at central regeneration facility. They earn their livelihoods from the regeneration activities and other technical services. These employment generations are a part of the long-term benefit generated by the arsenic remediation program. Apart from these direct employments, a host of secondary employments is generated such as water vendors who earn their livelihood by transporting the treated water to the households who live at far-off place or are unable to physically come to collect the treated water. Figure 10.10 shows a photograph of one such water vendor associated with a treatment unit. On average, each treatment unit has at least one such water vendor associated with it.

The communities with significant amount of accumulated fund are also investing the money to local infrastructure development programs. Some of the communities are also investing in providing credit to the farmers at soft interest rates and, also, in microcredit-based small businesses. Thus, the mitigation effort does not only become self-sustainable, but it also helps to turn the crisis into an opportunity for further economic growth.

Fig. 10.10 Photograph of a water vendor carrying arsenic-safe water to the households in West Bengal, India

10.9 Conclusion

This chapter discussed how small capital and appropriate technical know-how, when infused within the crisis-ridden areas in a proper way, went a long way in benefitting arsenic-affected villagers who are also stakeholders of each unit. The entire system involving the treatment of arsenic-contaminated water has eventually evolved as a viable business model where everybody associated got benefitted. Thus, the overwhelming crisis could be turned into an economic engine for further growth.

All these were possible due to a successful community participation program through which people learned about their rights. The community participation model that evolved over the time through the field-level experiences of the project team can be summarized of having four steps for development, namely, (i) external facilitation, (ii) skill and competency development, (iii) execution of self-assigned task, and (iv) expanding the village water committee (VWC) through creating linkage for sustainability. Figure 10.11 shows a flow diagram of such community participation model developed.

The essential criterion for community participation is the acceptance of the technology by the community which was possible due to the following reasons:

- The units did not change the way the water is collected; they were culturally appropriate, taking into account end-user lifestyles, preferences, etc.
- The cost of treated water produced is within the affordability limit of the villagers.
- The units are community level but are small enough to be owned, maintained, and operated by people residing in a village.

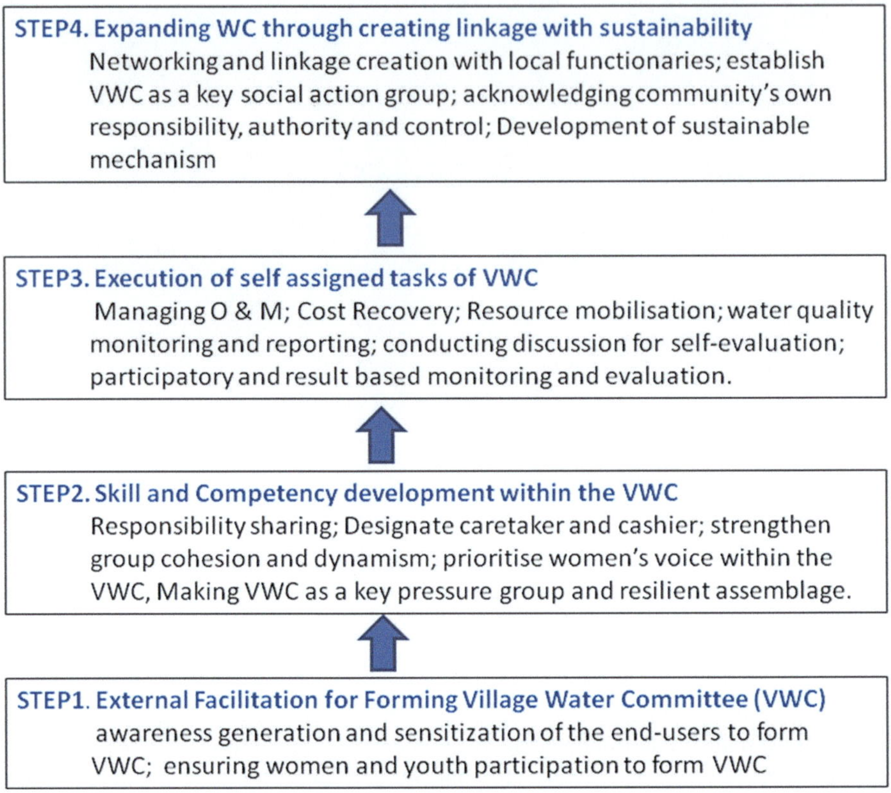

Fig. 10.11 Flow diagram for the development of the community participation model evolved in IIEST's arsenic mitigation project in West Bengal, India

- The units can be operated and maintained using locally available resource material by the villagers only with no or very little of training.
- The design is simple and flexible. It can be adapted to various places and changing circumstances.
- A small amount of capital is required.

The community not only actively took part in every stage in the arsenic mitigation projects; they also reached new heights by achieving success beyond mere sustainability. It was possible by involving into various types of developmental work for the society with the help of the savings that was made possible through effective collection and management of the water tariff. As these kinds of developmental activities grow over the time, the economic enterprise that grew out of the arsenic crisis shall gain more momentum so that it can fuel bigger changes in the society. If guided properly, the communities can, in long term, help promote practice of water-related hygiene, initiate sound sanitation practices, etc.

Acknowledgment The arsenic removal initiative that is discussed above was implemented with generous financial support from Water For People, USA. Effort of the members of SATHEE to run the central regeneration center and also the water testing facility is immensely valuable, without which the sustainability of the arsenic removal units would have been nearly impossible.

References

1. Abernathy CO, Liu YP, Longfellow D, Aposhian HV, Beck B, Fowler B et al (1999) Arsenic: health effects, mechanisms of actions, and research issues. Environ Health Perspect 107(7):593–597
2. Yoshida T, Yamauchi H, Sun FG (2004) Chronic health effects in people exposed to arsenic via the drinking water: dose-response relationships in review. Toxicol Appl Pharmacol 198(3):243–252
3. Lindberg AL, Rahman M, Persson LA, Vahte M (2008) The risk of arsenic induced skin lesions in Bangladeshi men and women is affected by arsenic metabolism and the age at first exposure. Toxicol Appl Pharmacol 230(1):9–16
4. Naujokas MF, Anderson B, Ahsan H, Aposhian HV, Graziano JH, Thompson C et al (2012) The broad scope of health effects from chronic arsenic exposure: update on a worldwide public health problem. Environ Health Perspect 121(3):295–303
5. Edwards M (1994) Chemistry of arsenic removal during coagulation and Fe-Mn oxidation. J Am Water Works Assoc 76:64–78
6. Ghosh MM, Yuan JR (1987) Adsorption of inorganic arsenic and organo-arsenicals on hydrous oxides. Environ Prog 3:150–157
7. DeMarco MJ, SenGupta AK, Greenleaf JE (2003) Arsenic removal using a polymeric/inorganic hybrid sorbent. Water Res 37:164–176
8. Driehaus W, Jekel M, Hildebrandt U (1998) Granular ferric hydroxide – a new adsorbent for the removal of arsenic from natural water. J Water SRT Aqua 47:30–35
9. Hering J, Chen PY, Wilkie JA, Elimelech M (1997) Arsenic removal from drinking water during coagulation. J Environ Eng ASCE 123:801–807
10. Min JM, Hering J (1998) Arsenate sorption by Fe(III)-doped alginate gels. Water Res 32:1544–1552
11. Ramana A, SenGupta AK (1992) Removing selenium (IV) and arsenic(V) oxyanions with tailored chelating polymers. J Environ Eng ASCE 118:755–775
12. Thirunavukkarasu O, Viraraghavan T, Subramanian K (2003) Arsenic removal from drinking water using iron oxide-coated sand. Water Air Soil Pollut 142:95–111
13. Zhang QJ, Pan BC, Chen XQ, Zhang WM, Pan BJ, Zhang QX, Lv L, Zhao XS (2008) Preparation of polymer-supported hydrated ferric oxide based on Donnan membrane effect and its application for arsenic removal. Sci China B 51:379–385
14. Lackovic JA, Nikolaidis NP, Dobbs GM (2000) Inorganic arsenic removal by zero-valent iron. Environ Eng Sci 17:29–39
15. Bajpai S, Chaudhury M (1999) Removal of arsenic from manganese dioxide coated sand. J Environ Eng ASCE 125:782–784
16. An B, Fu Z, Xiong Z, Zhao D, SenGupta AK (2010) Synthesis and characterization of a new class of polymeric ligand exchangers for selective removal of arsenate from drinking water. React Funct Polym 70:497–507
17. Dutta PK, Ray AK, Sharma VK, Millero FJ (2004) Adsorption of arsenate and arsenite on titanium dioxide suspensions. J Colloid Interface Sci 278:270–275
18. Driefhaus W, Seith R, Jekel M (1995) Oxidation of arsenic (III) with manganese oxides in water treatment. Water Res 29:297–305
19. Waypa JJ, Elimelech M, Hering J (1997) Arsenic removal by RO and NF membranes. J AmWater Works Assoc 89:102–114
20. Suzuki TM, Bomani JO, Matsunaga H, Yokoyama T (2000) Preparation of porous resin loaded with crystalline hydrous zirconium oxide and its application to the removal of arsenic. React Funct Polym 43:165–172

21. Su C, Puls RW (2008) Arsenate and arsenite sorption on magnetite: relations to groundwater arsenic treatment using zerovalent iron and natural attenuation. Water Air Soil Pollut 193:65–78
22. Ryu J, Choi W (2006) Photocatalytic oxidation of arsenite on TiO2: understanding the controversial oxidation mechanism involving superoxides and the effect of alternative electron acceptors. Environ Sci Technol 40:7034–7039
23. Sarkar S, Gupta A, Biswas RK, Deb AK, Greenleaf JE, Sen Gupta AK (2005) Well-head arsenic removal units in remote villages of Indian subcontinent: field results and performance evaluation. Water Res 39:2196–2206
24. Pena M, Meng X, Korfiatis GP, Jing C (2006) Adsorption mechanism of arsenic on nanocrystalline titanium dioxide. Environ Sci Technol 40:1257–1262
25. Ning RY (2002) Arsenic removal by reverse osmosis. Desalination 143:237–241
26. Hossain MA, SenGupta MK, Ahamed S, Rahman MM, Mondal D, Lodh D et al (2005) Ineffectiveness and poor reliability of arsenic removal plants in West Bengal, India. Environ Sci Technol 39(11):4300–4306
27. Ministry of Water Resources, Government of India, National Water Policy (2002, 1st April) New Delhi. Available at: http://mowr.gov.in/writereaddata/linkimages/nwp20025617515534.pdf. Accessed on 22 July 2013
28. Sarkar S, Blaney LM, Gupta A, Ghosh D, Sen Gupta AK (2008) Arsenic removal from groundwater and its safe containment in a rural environment: validation of a sustainable approach. Environ Sci Technol 42(12):4268–4273
29. Sarkar S, Blaney LM, Gupta A, Ghosh D, Sen Gupta AK (2007) Use of ArsenX^{np}, a hybrid anion exchanger, for arsenic removal in remote villages in the Indian subcontinent. React Funct Polym 67(12):1599–1611
30. Sarkar S, Greenleaf JE, Gupta A, Ghosh D, Blaney LM, Bandyopadhyay P et al (2010) Evolution of community-based arsenic removal systems in remote villages in West Bengal, India: assessment of decade-long operation. Environ Sci Technol 44(19):5813–5822

Anirban Gupta has 25 years of teaching and research experience. Presently, he is serving as a professor of civil engineering and as the associate dean at IIEST, Shibpur. His research is focused on water treatment, specially removal of contaminants like arsenic and fluoride from drinking water. He has several publications in international books and journals. He has been active in social outreach of the research initiatives and installed more than 150 arsenic and fluoride removal systems in more than 200 villages and schools in several states in India. His involvement with NGOs helped to provide water and sanitation solutions to many villages. He is a member of the Arsenic Task Force, Government of West Bengal and the Director of Rural Development Forum, Institution of Engineers (India). He is a Fulbright Fellow. He has to his credit awards from UNESCO, USEPA, ASCE, IChemE-UK and Bharat Chamber of Commerce.

Dr. Sudipta Sarkar is an assistant professor of environmental engineering within the departmental of civil engineering at Indian Institute of Technology (IIT) Roorkee. He earned his PhD from Indian Institute of Engineering Science and Technology (IIEST), Shibpur in the year 2006. Prior to joining IIT Roorkee, he spent about 6 years at Lehigh University, USA as a postdoctoral research scholar. During his sojourn with IIEST, Dr Sarkar has played an active role in research, development and implementation of about 150 community-scale arsenic removal devices. His research interest and activity spans across various areas of water, wastewater engineering and solid waste management, including environmental nanotechnology. Before switching over to the academics, he has worked in core industry for about 10 years when he was engaged in developing designs for water and wastewater systems for municipalities as well as industries. He has authored about 20 research articles in journals of repute, US patents and several book chapters. Although his work is well cited, and he received many awards and accolades, he considers that the best reward he ever received is the reflections of gratitude in the eyes of the arsenic-survivors who now use to drink safe water from the arsenic removal units he has helped to install.

Dr. Debabrata Ghosh was attached with arsenic removal project of Indian Institute of Engineering Science and Technology (IIEST), Shibpur from 2004 to 2008. He obtained his Masters Degree in Environmental Engineering in 2002 and PhD in Civil Engineering in 2011 from the same Institute. During his work in arsenic removal project as Project Manager, he supervised installation, testing and monitoring of over 100 community level arsenic removal units. Furthermore he developed project procedures, monitoring methods, regeneration protocols, central regeneration facilities etc. Before joining arsenic removal project of IIEST, he worked in different industries for 35 years after passing B.Sc. from Calcutta University and Associate Membership examination from The Indian Institute of Metals. He has authored 10 research articles in international and national journals of repute. He was one of the recipients of P3 award offered by United States Environmental Protection Agency.

Dr. Kabita Maiti has training on community health and she is engaged in community mobilisation with different NGOs for empowerment on safe drinking water and sanitation since 1994. The main role was to initiate and develop a social process within the target communities for collective analysis of community needs, collective action leading to solution of the problem and to make the process self-sustaining and self-managed within the beneficiary community. She has experience for working with communities for arsenic and fluoride mitigation in West Bengal, UP, Bihar and Karnataka. Presently she is working as project coordinator in SATHEE, an NGO in West Bengal.

A Cost-Effective Technology for Arsenic Removal: Case Study of Zerovalent Iron-Based IIT Bombay Arsenic Filter in West Bengal

11

Tuhin Banerji and Sanjeev Chaudhari

11.1 Introduction

Arsenic contamination in drinking water has affected people all over the world. Arsenic poisoning from contaminated drinking water is an especially serious problem in parts of West Bengal (India) and most of Bangladesh. Arsenic contamination in water is of grave concern in Indian perspective. The Bengal Delta plain is drained by three important rivers, the Ganges, Meghna and Brahmaputra (G-M-B), which originate from the Himalayas. Therefore, the Bay of Bengal receives maximum amount of sediments from G-M-B river system containing several trace elements including arsenic [1].

Arsenic concentrations in sediments from the river Ganges averaging 2.0 mg kg^{-1} (range 1.2–2.6 mg kg^{-1}), from the river Brahmaputra averaging 2.8 mg kg^{-1} (range 1.4–5.9 mg kg^{-1}) and from the river Meghna averaging 3.5 mg kg^{-1} (range 1.3–5.6 mg kg^{-1}) have been found [2]. Arsenic contamination in groundwater was also detected in Rajnandgaon district of Chhattisgarh and Ballia district of Uttar Pradesh [3]. Apart from West Bengal and Gangetic plain of Uttar Pradesh, the states of Bihar, Jharkhand and Brahmaputra plain in Assam and other northeastern states and some regions of Punjab, Haryana, Himachal Pradesh, Delhi, Chandigarh and Rajasthan also have elevated levels of arsenic in their groundwater. At concentrations found in natural waters, arsenic is the most carcinogenic of all substances named in current drinking water regulations [4]. Recent studies also indicate additional contamination of the food chain due to irrigation with arsenic-contaminated water [5, 6]. Some years back, World Health Organization (WHO) lowered the guideline value for arsenic in drinking water from 50 to 10 ppb [7]. In the past 5 years, many countries, including India, have adopted 10 ppb as the maximum

T. Banerji (✉) • S. Chaudhari
Centre for Environmental Science and Engineering, Indian Institute of Technology Bombay, Powai, Mumbai 400076, India
e-mail: t.banerji@neeri.res.in; sanjeev@iitb.ac.in

K.J. Nath, V.P. Sharma (eds.), *Water and Sanitation in the New Millennium*,
DOI 10.1007/978-81-322-3745-7_11

Fig. 11.1 Groundwater arsenic contamination of G-M-B plain in India ("●" mark the areas of greatest concern)

contamination level (MCL) in drinking water. A lower limit of 2 μg/L was suggested for a better safety margin but rejected due to the financial implications and lack of available technologies [8]. Figure 11.1 enlists the areas in India where arsenic has been detected over acceptable limits as per Indian standards for drinking water.

The installation of tube wells in an effort to provide easy accessibility to drinking water free of microbial pathogens has resulted in widespread arsenic poisoning of people living in these areas [9, 10]. The use of arsenic-free water source is one of the best solutions to get rid of arsenic poisoning-related problems. But this is not always possible. The western countries generally prefer centralised water treatment systems where proper control and monitoring can be done. But in India due to the remoteness of villages and the costs involved in using centralised schemes, the best option would be to use many small decentralised water treatment units. Various treatment technologies have been developed for arsenic removal from drinking water. The commonly used technologies include coagulation and precipitation with iron and aluminium salts, adsorption of natural ores or granular ferric hydroxide (GFH) or activation of alumina, ion exchange and reverse osmosis. Coagulation-precipitation and adsorption techniques with activated alumina and ferric hydroxide have been found to be less efficient for As(III) removal than As(V) removal. As groundwater usually contains an appreciable proportion of As(III), an efficient arsenic removal by most of the technologies necessitates conversion of As(III) to As(V) by chemical oxidation. In general, the available technologies for arsenic removal are uneconomical and require skilled manpower for operation, and also none of the technologies claim to be able to reach 10 ppb level of arsenic in drinking water consistently. Also some technologies like the GFH are patented so they are very expensive. Others which use alumina leach a significant quantity of aluminium in the treated water, but the monitoring of Al^{3+} leaching out is not done. Aluminium is a known neurotoxin, and such technologies add the danger of aluminium into the drinking water system. Another point of concern for the presently available technologies is that as they involve the use of adsorbents, the adsorbents have a limited life after which they are saturated. These adsorbents then need to be either regenerated or replaced. This requires the presence of a supply chain which will increase costs, and such supply chains may not be accessible to the remote areas.

In spite of increased public awareness and concern, treated arsenic-free drinking water has not reached the common mass. In developing countries, the cost incurred towards installation of treatment plants and their operation and maintenance has been a major issue [11]. Treatment plants installed and operated by public authorities are generally large and located in centralised places, and distribution of treated water to each and every user is a point of concern. Many of the researchers are therefore interested in developing low-cost treatment facilities like domestic filters, which can be installed at the users' locations. Secondly, provision of low-cost and easily available adsorbents has also been looked into with an objective to reduce the operation costs.

But small domestic units are very difficult to maintain as the maintenance of the units has to be left to the villagers themselves, and specialised supervision will be difficult. There are reports that even the villagers find that the community filters are more suitable to them [12–15]. Villagers have reported a preference for fetching water from deep tube wells over using and maintaining a household-based filter to remove arsenic, but their most common complaint is the distance of travel to the deep tube well [14]. Most villagers have shallow (and inexpensive) tube wells in their yard. In several villages in the district of Jessore, in Bangladesh, villagers who had to walk 200–500 m to fetch water from community-owned deep tube wells considered this distance to be far. While people reported that they generally fetched drinking water from the deep tube well, most villagers also admitted that they would drink water coming out of contaminated tube wells when they did not want to make the trip to the arsenic-free source.

Caldwell et al. [16] explains that fetching water from outside family compounds and then maintaining a household water treatment unit increase the workload of Bangladeshi women. They would rather use a one-step process, i.e. bring water from a distance and then forget about it. If they are asked to treat it further through domestic filters, they find it to be a cumbersome task. Therefore, minimising the distance to safe water sources is important for the success of a community water treatment unit.

Recently, zerovalent iron (ZVI) has been actively investigated as a viable alternative for removal of arsenic from groundwater [5, 17–20]. Several researches have also been carried out by using ZVI in the domestic filter for arsenic removal [21, 22]. In such filters, sand was used as the filter media. However, clogging of the filter bed was reported during longer operations [22–24].

Selection of a suitable method to supply arsenic-free drinking water depends on several factors. In the context of Bengal delta, where centralised water supply is not available, the factors to be considered for selecting an appropriate solution for the supply of potable water are:

- Simple to operate (should not require specially trained personnel)
- Robust (should not break down often and give consistent performance)
- Low cost (cost of treatment should be low as the villagers have low average income)
- Able to function without electricity
- Based on local resources and skills
- Accessible to community and women's groups or the village elders
- Include provisions for a safe method of disposal of As-rich sludge

But over and above all these factors, in the rural communities of Bengal, costs and perceived convenience are most important in increasing technology adoption rates. Examination of relevant observations and informal interviews will lead to the development of design strategies that ensure technologies and implementation plans will be acceptable to users. Only when all the concerns of the villagers are addressed specifically will a possible implementation plan emerge. Thus, a technology was needed which would use only material available locally at the village and was so simple and cheap that an unskilled person could operate and maintain it.

Thus, this work was started with the goal of developing a robust community scale arsenic removal technology for rural population. This method for arsenic removal is based on iron nail (ZVI) corrosion which in the presence of dissolved oxygen co-oxidises As(III) to As(V). Subsequently As(V) formed adsorbs onto the hydrous ferric oxide (HFO) formed from the corrosion of ZVI. In this technology, the iron requirement is 20 times less for similar arsenic removal efficiency than what is reported in the literature. Thus, arsenic removal is achieved without any proprietary chemicals.

11.2 Performance of IITB Arsenic Filter

During January 2008, Indian Institute of Technology Bombay (IITB) installed one Arsenic Removal Plant (ARP) at Sonakhali Mouza (village Shikaripara Sonakhali) of Nadia district. The first plant was installed in a shallow tube well of 50 m depth. In the same year in the month of February, another ARP was installed in Ganguria Mouza (village Ganguria) of Haringhata block of Nadia district. This tube well was also of shallow depth (52.5 m). In October 2009, IIT Bombay installed two more ARPs, one in Palashi Mouza (village Palashi Ranipara) of Barrackpur I block of North 24 Parganas district and the other in Sonakhali Mouza (village Sonakhali Biswaspara) of Haringhata block of Nadia district. Both the plants were installed in shallow tube wells (39 m and 33 m, respectively). Table 11.1 shows the details about the location and the installation date of the IITB arsenic filters.

The arsenic filter can be attached to lift-and-force handpumps, and so it can be installed on any pre-existing borehole. The filter unit in this system consists of two tanks which are modular. Each tank has a section on top where the water is contacted with iron nails (ZVI). After ZVI contact, the water trickles down to the filter

Table 11.1 Location and the installation date of the IITB arsenic filters in West Bengal

Plant no.	Address	Latitude and longitude	Depth of tube well (m)	Age of tube well at the time of installation of ARP	Date of installation of ARP
1	Village, Sonakhali Biswaspara; Mouza, Sonakhali; GP, Birohi; Block, Haringhata; Dist, Nadia	N 23° 00′ 16.1″ E 88° 33′ 08.3″	33	11 years	16 October 2009
2	Village, Sonakhali Shikaripara; Mouza, Sonakhali; GP, Birohi; Block, Haringhata; Dist, Nadia	N 22° 59′ 59.5″ E 88° 33′ 16.0″	50	10 years	27 January 2008
3	Village, Ganguria; Mouza, Ganguria; GP, Haringhata I; Block, Haringhata; Dist, Nadia	N 22° 57′ 29.2″ E 88° 33′ 17.0″	52.5	12 years	02 February 2008
4	Village, Palasi Ranipara; Mouza, Palashi; GP, Palashi Majipara; Block, Barrackpur I; Dist, North 24 Parganas	N 22° 55′ 08.2″ E 88° 29′ 25.8″	40	8 years	18 October 2009

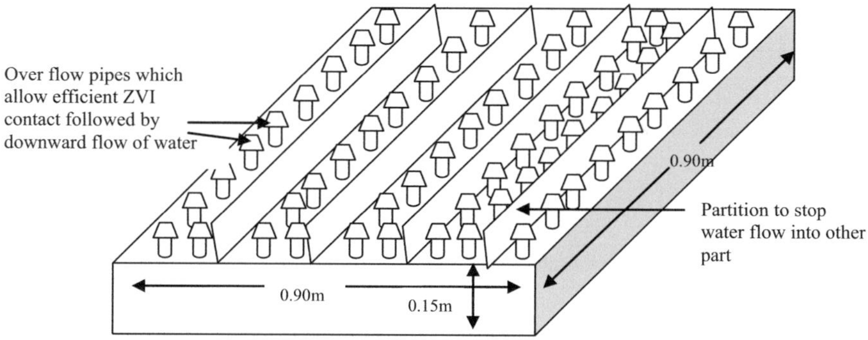

Fig. 11.2 Design details of specially designed ZVI contactor tray

bed, and in this process, the oxidation of Fe^{2+} and As(III) takes place. The filter bed is made up of stone aggregates used for building construction.

The IITB arsenic filter consists of a ZVI contactor followed by a gravel filter, both of these are housed in the same tank (the ZVI contactor tray has 1" non-galvanised iron nails and is placed on a tray support). The ZVI contactor tray design details are shown in Fig. 11.2. It comprises a square tray of 0.9 m × 0.9 m with overflow pipes covering its entire area. The bottom of the tray is drilled, and the overflow pipes are welded to it. The overflow pipe design is shown in Fig. 11.3. Each overflow pipe is a hollow pipe of diameter 2.5 cm and height of 11.5 cm. The

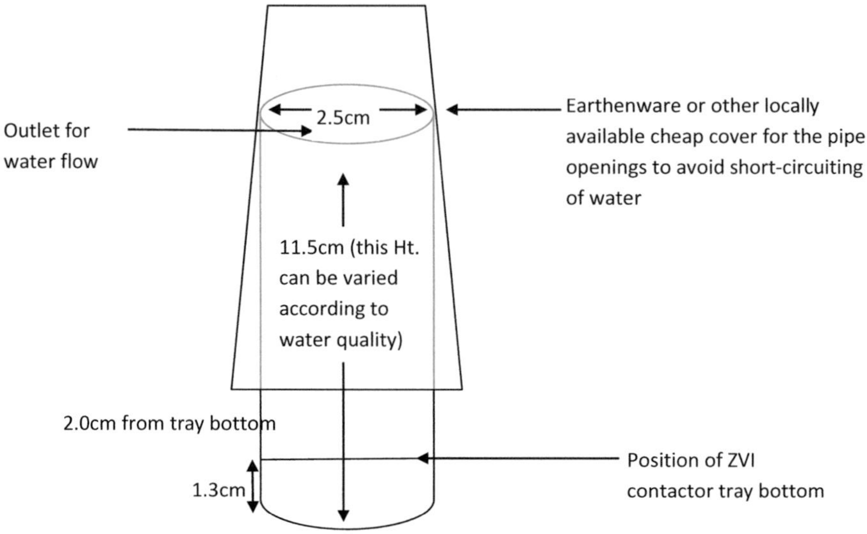

Fig. 11.3 Design details of the overflow pipe in the ZVI contactor tray

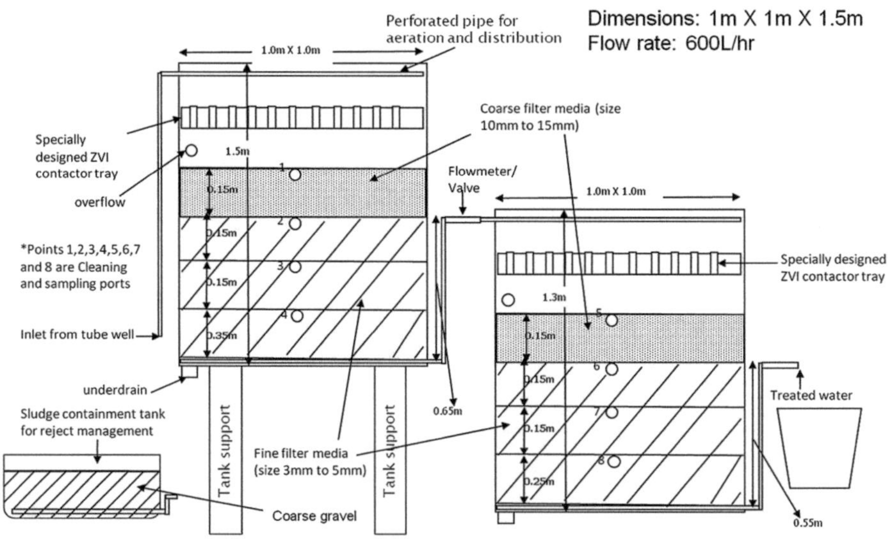

Fig. 11.4 Schematic of IITB arsenic filter installed at the test sites in West Bengal

height and number of overflow pipes may vary as per the influent water flow. The top opening of the overflow pipe is covered with conical cover, as per the design. This cover should not be higher than 2.0 cm from the tray bottom. The lower 1.3 cm of the overflow pipe is below the tray. All the welds and joints in the tray are made leakproof. The filtration zone for separation of hydrous ferric oxide (HFO) floccules is shown in Fig. 11.4. This is the design of the pilot plant installed in the test sites in

West Bengal. The filtration chamber consists of a tank of dimensions 1 m × 1 m × 1.4 m. The tank is filled with gravel and aggregates of the size of 2–4 mm up to the height of 0.65 m. Then another 0.15 m is filled with gravel and aggregates of the size 4–7 mm. The top 0.6 m is left free for ZVI unit and inlet pipe. Cleaning ports are present at a distance of 0.65 m, 0.50 m and 0.35 m from the bottom, and an overflow is provided at 0.80 m from the bottom. Another filter tank with similar configuration is set up along with the first tank. This is for double-stage filtration, as an added fail-safe mechanism. The arsenic removal plants were first installed at Shikaripara and Ghanguria, both in Nadia district of West Bengal, on 27 January 2008 and 2 February 2008, respectively. Since then the maintenance of the plants has been done by the local villagers.

11.2.1 Pilot Plant Performance

In this process, due to the leaching of iron when in contact with water, iron (Fe^0) gets dissolved in the form of Fe^{2+}, and under the mild oxidising environment, Fe^{2+} gets oxidised to Fe^{3+} forming high oxidising intermediates which co-oxidises As(III) to As(V). Hence, As(III) gets oxidised without addition of chemical oxidant. Further Fe^{3+} forms insoluble hydrous ferric oxide (HFO) in the pH range of 6.5–8.3 (which is the pH range of groundwater) which adsorbs arsenate [HFO has quite high affinity for As(V)] and is separated by the designed filtration chamber. The process developed is quite simple and easy to operate and achieves complete oxidation of arsenite to arsenate. The process is able to achieve arsenic levels below 10 ppb for initial arsenic concentrations ranging from 300 to 1000 ppb. The reactor is able to consistently provide drinking water for 100–200 families at all sites on a daily basis and can be scaled up to cater to much larger communities. The IITB arsenic filter requires cleaning once in 3 months, and it is done by the villagers; therefore, it has low operation and maintenance costs. Moreover, it does not require monitoring of flow parameters and is easy to operate by unskilled personnel. Design flow rate was 600 L/h, but it has been operated at times at 1000 L/h. at all sites; even then the treated water arsenic concentration is not seen to increase.

For monitoring the plants, fortnightly sample collection was carried out from each site. On site testing was done with the Wagtech Digital Arsenator (instrument manufactured by Wagtech International, UK), and then these samples were acid preserved and couriered to IIT Bombay for detailed analysis. The analysis of the samples at IIT Bombay was carried out by molybdenum blue method developed by Dhar et al. [25] and the FIAS-AAS (flow injection analysis system for atomic absorption spectrophotometer) (Perkin Elmer model 400AA with FIAS 100). Samples were also analysed by local accredited labs using Gutzeit method.

The results of this monitoring have been summarised in Table 11.2. From the results, it can be seen that the filters are able to provide water of low arsenic concentrations at each of the sites. The annual average of treated water is less than or around 10 ppb in most cases. At times the reading has been observed to be above the

Table 11.2 Results of fortnightly monitoring of the ARPs at the four *test* sites in West Bengal

Location	Sonakhali Biswaspara	Sonakhali Shikaripara	Ghanguria	Palasi Ranipara
Average annual initial arsenic concentration (ppm)	0.2615 (±0.0264)	0.1674 (±0.0232)	0.3881 (±0.0400)	0.0715 (±0.0128)
Average arsenic concentration in treated water in 2010 (ppm)	0.0061 (±0.0034)	0.0090 (±0.0034)	0.0065 (±0.0031)	0.0108 (±0.0025)
Average arsenic concentration in treated water in 2011 (ppm)	0.0069 (±0.0037)	0.0081 (±0.0051)	0.0108 (±0.0154)	0.0090 (±0.0043)
Average arsenic concentration in treated water in 2012 (ppm)	0.0082 (±0.0057)	0.0136 (±0.0072)	0.0188 (±0.0102)	0.0102 (±0.0051)

WHO limits, it is because of poor maintenance of the units by the villagers. That is why some of the averages are slightly on the higher side.

The arsenic removal plant design based on the ZVI is an economical alternative to the existing technologies, and further it meets the WHO drinking water guideline value of 10 μg/L. Also the sludge generation from these plants is minimal, thus the problem of arsenic-laden sludge is also minimised. Other projects have very inefficient sludge management systems, but for this process, sludge collection and containment system have been designed and installed which can hold the sludge for more than 5 years, without arsenic leaching out from it.

All the available technologies using iron-based adsorbents require a ratio of Fe/As of 250 or more to achieve low arsenic concentrations in treated water. Fe/As ratio is the ratio of iron added to the amount of arsenic removed. This gives us an idea about the amount of sludge generated by the system. High Fe/As ratios mean more iron requirement and more sludge generated which translates to higher costs for sludge disposal and cumbersome maintenance. The IITB arsenic filter uses a Fe/As ratio of 10. That is why the cost of the unit is so low and easy to maintain.

The pilot plants were modified extensively using the feedback from the villagers and made more suitable to rural India. Thus, four IITB arsenic filters have been installed and are operational in four villages of West Bengal, India, since 2008. The highest average arsenic concentration in the raw water being treated by the IITB arsenic filter in West Bengal is 0.4 mg/L in the village of Ghanguria in Nadia District. The average arsenic concentration in the filtered water is less than 0.01 mg/L. The IITB arsenic filter seems to have promise for use in rural India. Apart from the units mentioned previously, 56 more IITB arsenic filters now are operational in Uttar Pradesh (27 nos.), Bihar (21 nos.), Assam (4 nos.) and West Bengal (4 nos. more other than those mentioned). The cost for fabrication of each IITB arsenic filter varies from Rs. 40,000 to 75,000 depending on the cost of material at different places and cost of transportation. The annual maintenance costs including the cost of maintenance of the handpump, cleaning the filter and replenishing the iron, are around Rs. 1000.

Independent evaluation of the IITB arsenic filter has been carried out by the School of Environmental Science, Jadavpur University, West Bengal, for a period of 8 months from January to August 2010. They found that the ARPs were efficient in filtering water below 20 μg/L consistently, but none of these plants was able to remove arsenic to less than 10 μg/L except on a few occasions. Subsequently the pilot plants were further modified, and later analysis done by the same laboratory found that all the units had arsenic of less than 10 μg/L in the treated water.

11.3 Conclusions

The arsenic removal filter developed by IITB has the following salient features:

- IITB arsenic removal filter uses only locally available materials and constructed by local plumbers and masons.
- The process achieves oxidation of As(III) to As(V) and subsequent arsenic removal by hydrous ferric oxide, which is formed from oxidation of leached Fe^{2+}, without the addition of any chemicals.
- The unit is cost-effective and robust and does not require extensive monitoring.
- The unit is simple to operate and requires less maintenance.
- Twenty times less sludge is generated, as compared to current technologies.
- The unit has a simple design which enables local persons to replicate it.
- IITB arsenic filter is designed to provide drinking water to meet the daily needs of 300 families.
- The cost for producing 1 cubic metre of water (average monthly requirement for a family of five persons) varies from Rs. 0.10 to Rs. 1.00.

Thus, due to all these features and successful testing phase operation in the field, it is felt that the IITB arsenic filter is suitable as an arsenic removal filter for rural India.

Acknowledgements The authors would like to thank the Department of Science and Technology, Government of India, for funding the research work.

References

1. Mukherjee A, Sengupta MK, Hossain MA, Ahamed S, Das B, Nayak B et al (2006) Arsenic contamination in groundwater: a global perspective with emphasis on the Asian scenario. J Health Popul Nutr 24(2):142–163
2. Datta DK, Subramanian V (1997) Texture and mineralogy of sediments from the Ganges-Brahmaputra-Meghna river system in the Bengal Basin, Bangladesh and their environmental implications. Environ Geol 30(3–4):181–188
3. Ahamed S, Sengupta MK, Mukherjee A, Hossain MA, Das B, Nayak B et al (2006) Arsenic groundwater contamination and its health effects in the state of Uttar Pradesh (UP) in upper and middle Ganga plain, India: a severe danger. Sci Total Environ 370(2–3):310–322

4. Reimanna C, Matschullatb J, Birkec M, Salminend R (2009) Arsenic distribution in the environment: the effects of scale. Appl Geochem 24(7):1147–1167
5. Manning BA, Hunt M, Amrhein C, Yarmoff JA (2002) Arsenic (III) and arsenic (V) reactions with zero valent iron corrosion products. Environ Sci Technol 36:5455–5461
6. Huq SMI, Joardar JC, Parvin S, Correll R, Naidu R (2006) Arsenic contamination in food-chain: transfer of arsenic into food materials through groundwater irrigation. J Health Popul Nutr 24(3):305–316
7. WHO (2008) Guidelines for drinking-water quality: incorporating 1st and 2nd addenda, Vol. 1, Recommendations. – 3rd ed., WHO Library Cataloguing-in-Publication Data, Geneva. Available at http://www.who.int/water_sanitation_health/dwq/gdwq3rev/en/. Accessed on 27 July 2008
8. Smith AH, Lopipero PA, Bates MN, Steinmaus CM (2002) Arsenic epidemiology and drinking water standards. Science 296:2145–2146
9. Bhattacharjee Y (2007) A Sluggish response to humanity's biggest mass poisoning. Science 315(5819):1659–1661
10. Halsey PM (2000) Arsenic contamination study of drinking water in Nepal, ME thesis submitted to Civil and Environmental Engineering, Massachusetts Institute of Technology, USA
11. Reisinger HJ, Burris DR, Hering JG (2005) Remediation subsurface arsenic contamination with monitored natural attenuation. Environ. Sci. Technol 39(22):458A–464A
12. Mathieu J L, Gadgil A J, Kowolik K, Addy SEA (2008) Removing arsenic from contaminated drinking water in rural Bangladesh: Recent fieldwork results & policy implications. 2008 UNC Environmental Symposium Safe and Sustainable Drinking Water in Developing and Developed Countries: Where Science Meets Policy. Available at http://arsenic.lbl.gov/pdf/unc-symposium-aruba.pdf. Accessed on 25 July 2010
13. Mathieu JL, Gadgil AJ, Kowolik K, Qazi S, Agogino AM (2010) Design strategies and preliminary prototype for a low-cost arsenic removal system for rural Bangladesh. Lawrence Berkeley National Laboratory: Lawrence Berkeley National Laboratory. LBNL paper LBNL-2696E. Available at http://www.escholarship.org/uc/item/6db8c8vn. Accessed on 25 July 2010
14. Ahmad J, Goldar BN, Misra S, Jakariya M (2003) Willingness to pay for arsenic-free, safe drinking water in Bangladesh, World Bank Water and Sanitation Program- South Asia. Available at http://www.cepis.org.pe/bvsacd/arsenico/willing.pdf. Accessed on 25 July 2010
15. Ahmad J, Goldar B, Misra S (2005) Value of arsenic-free drinking water to rural households in Bangladesh. J. Environ. Manage 74(2):173–118
16. Caldwell BK, Caldwell JC, Mittra SN, Smith W (2003) Searching for an optimum solution to the Bangladesh arsenic crisis. Soc Sci Med 56:2089–2096
17. Farrell J, Wang JP, O'Day P, Conklin M (2001) Electrochemical and spectroscopic study of arsenate removal from water using zero valent iron media. Environ Sci Technol 35(10):2026–2032
18. Melitas N, Wang J, Conklin M, O'Day P, Farrell J (2002) Understanding soluble arsenate removal kinetics by zero valent iron media. Environ Sci Technol 36:2074–2081
19. Nikolaidis NP, Dobbs GM, Lackovic JA (2003) Arsenic removal by zero-valent iron: field, laboratory and modeling studies. Water Res 37:1417–1425
20. Bang S, Korfiatis GP, Meng X (2005) Removal of arsenic from water by zero valent iron. J Hazard Mater 121(1–3):61–67
21. Ngai TKK, Shrestha RR, Dangol B, Maharjan M, Murcott SE (2007) Design for sustainable development- Household drinking water filter for arsenic and pathogen treatment in Nepal. J Environ Sci Health-Part A 42(12):1879–1888
22. Mehta VS (2008) Arsenic removal from drinking water using Zero Valent Iron (ZVI). M. Tech. dissertation, Centre for Environmental Science and Engineering, Indian Institute of Technology Bombay
23. Sutherland D, Swash PM, MacQueen AC, McWilliam LE, Ross DJ, Wood SC (2002) A field based evaluation of household arsenic removal technologies for the treatment of drinking water. Environ Technol 23(12):1385–1404

24. Leupin OX, Hug SJ, Badruzzaman ABM (2005) Arsenic removal from Bangladesh tube well water with filter columns containing zero valent iron filings and sand. Environ Sci Technol 39(20):8032–8037
25. Dhar RK (2004) A rapid colorimetric method for measuring arsenic concentrations in groundwater. Anal Chim Acta 526:203–209

Dr. Tuhin Banerji has worked on various aspects of water and wastewater treatment technologies. He completed his PhD from IIT Bombay on arsenic removal using naturally present groundwater iron. During his PhD, he developed an arsenic removal filter which was suitable for rural areas of India, and it was very well received by all villagers wherever tested. Post PhD, he joined CSIR-NEERI and has been working on various wastewater-related issues.

Dr. Sanjeev Chaudhari is professor at the Centre for Environmental Science and Engineering, at Indian Institute of Technology Bombay. He earned his MTech from IIT Kanpur and PG diploma from Delft University. He completed his PhD from IIT Kanpur. Prof. Chaudhari specializes in water and wastewater treatment technologies and has developed zero-valent iron-based community-scale arsenic removal filter for rural India. Prof. Chaudhari was awarded with the Best Young Scientist Award in the 2nd M.P. Young Scientist Congress, Bhopal, in 1987 and Career Award for Young Teachers in the year 1996. Prof. Chaudhari is a reviewer for many prominent journals and has reviewed several research projects for DST and CSIR.

Urban Solid Waste Management: Key Issues and Challenges

12

Arunabha Majumder

12.1 Introduction

Most of the cities and towns in India are plagued by the acute problems related to solid waste. It is estimated that total municipal solid waste generated by 300 million urban populations is around 44 million tons per year. Municipal authorities provide conservancy services to the citizen, and accordingly they are responsible for collection, transportation, treatment, and disposal of solid waste. Unfortunately, conservancy services extended by the municipalities are mostly poor resulting in irregular collection, accumulation, and uncontrolled dumping of solid waste. Practices of throwing solid wastes along footpaths, roads, curb channels, or drains are very common in most of the urban centers. Such practices along with uncontrolled open dumping of solid wastes result in surface and groundwater pollution, odor problem, fly/mosquito/insect nuisance, and air pollution. Mismanagement in tackling municipal solid waste often develops insanitary condition of urban localities and causes serious threat to public health and environment.

Components of solid waste management (SWM) include source segregation and household storage, collection, transfer and transportation, treatment, recycling, and disposal. These components are interlinked and require sound operation and management so that generated wastes do not cause any deleterious effect on public health and environment.

Municipal solid waste is heterogeneous in nature comprising of organic and inorganic components. A part of the inorganic wastes is recyclable, and as such, those are salvaged by ragpickers/scavengers for their livelihood. Inorganic recyclables salvaging in unorganized manner and use the same in recycling units may pose risk and threat to cooupational health as well as degradation of environment.

A. Majumder (✉)
Emeritus Professor, School of Water Resources Engineering, Jadavpur University,
Kolkata 700032, India
e-mail: arunabhamajumder@hotmail.com

K.J. Nath, V.P. Sharma (eds.), *Water and Sanitation in the New Millennium*,
DOI 10.1007/978-81-322-3745-7_12

Thus, recyclable waste segregation and their use in recycling units require institutional development and capacity building for socioeconomic benefit of scavengers or ragpickers and also toward supporting to achieve sustainability to solid waste management.

In municipal solid waste, a considerable quantum is organic which can be converted to compost through biological process. The compost is suitable to be used as soil conditioner having total (average) nitrogen, phosphorus, and potassium content of 3% by weight. But except a few, successful composting units in cities and towns are yet to be established. Performances of many compost plants are not satisfactory, and as such, sustainability of compost units appears to be a serious challenge to the city managers.

Disposal of inorganic nonrecyclable solid waste can be disposed by secured (sanitary) landfilling process subject to availability of land. Most of the municipalities do not have facilities for disposal of the solid waste in secured landfill site; instead they practice uncontrolled disposal of solid wastes by open dumping. Again, many municipalities are facing a problem for disposal of solid wastes because of land crunch and constraint.

Resource recovery through waste processing can result in recovery of useful products such as compost and energy. Waste to energy process may lead to generate electricity. Generation of refused derived fuel (RDF) may be a good option in managing municipal solid waste. Resource recovery through waste processing may partially solve the future land requirement problem of the municipalities. But in this context, it is essential to examine the technical feasibility of different resource recovery processes. Basic constraint of conversion of waste to energy is the low calorific value (1100–1400 Kcal/kg), higher moisture content (60–70%) and the presence of even 20–25% of inert matters. Today, it is encouraging to note that many municipal authorities are seriously thinking on the feasibility of converting waste to energy for deriving a sustainable solution for urban centers of the country.

The Ministry of Environment and Forest, Govt. of India, has framed Municipal Solid Wastes (Management and Handling) Rules 2000 as per provision in Environment (Protection) Act 1986, where it is stated that every municipal authority shall, within the territorial area of the municipality, be responsible for the implementation of the provision of these rules and for any infrastructure development for collection, storage, separation, transportation, processing, and disposal of solid waste. It is the responsibility of all urban local bodies in the country to prepare long-term plans for effective solid waste management in their respective cities and towns. Uncontrolled open dumping for disposal of solid waste must be discontinued, and all options of processing, treatment, and resource recovery should be studied to select appropriate option for managing municipal solid waste. A challenge to the urban local bodies for selecting appropriate technology processes to achieve perfection in operation and maintenance of the system in a suitable way is of no doubt.

Table 12.1 Composition and characteristics of municipal solid waste

Composition	(% by weight)
Garbage	45–55
Ash/earth inert	20–25
Paper/cardboard box	5–6
Plastic/PVC/HDPE	6–7
Rag/cotton/textile	1–3
Rubber	0.3–0.5
Leather	0.08–1.0
Earthenware and ceramics	3–4
Glass	1–2.5
Hay/straw/thermocol	0.1–0.2
Leaves	2–4
Metal	0.25–0.5
Others	4–5

12.2 Waste Generation and Characteristics

The generation and characteristics of municipal solid waste are influenced by the socioeconomic condition, available conservancy services, climate, social development, and cultural practices. It is always advisable to carry out qualitative and quantitative analysis of municipal solid waste for undertaking long-term planning of solid waste management system for urban local bodies. However, a general pattern of composition and characteristics of municipal solid waste is presented in Table 12.1.

The generation of solid waste in cities and towns (India) ranges between 350 gms/cap/day and 600 gms/cap/day. It has varying density ranging between 300 and 500 kg/cub.m. The moisture content of solid waste normally ranges between 50 and 70%.

In most of the cities/towns, solid wastes are not segregated at the source of generation. Certain specific items of wastes are segregated at the household level for selling. In addition, around 15% of wastes are salvaged by the scavengers/ragpickers from different locations including disposal ground. Thus, characteristics of solid waste get slightly changed from the source of generation to the disposal site.

12.3 Components of Solid Waste Management

The following are the components of municipal solid waste management:

- Source segregation and storage
- Collection
- Transportation
- Recycling
- Treatment
- Disposal

All the components of SWM are interlinked, and failure in the activity of one component will result in a setback of the whole management system. Each of the components of SWM must be planned according to the existing situation and condition of the city/town so that operational SWM system becomes sustainable.

12.4 Status of Urban Solid Waste Management

The status of SWM in most of the ULBs in India can be highlighted as follows:

- The generation of solid waste (per capita basis) increases with the increase in density of population in the municipal towns.
- The generation of solid waste ranges between 300 and 650 gms/cap/day.
- A considerable portion of solid waste gets accumulated in different parts of the town every day. The collection of solid waste ranges between 80 and 93% per day.
- Solid waste collection frequency is either daily or alternately.
- Solid waste collection from house to house, roadside vats, community vats, market vats, etc. is practiced.
- Solid wastes are mostly not segregated at the source of generation.
- The primary collection system in most of the municipal towns needs improvement.
- Primary transfer stations (ward wise) are yet to be developed with proper infrastructure in most of the towns.
- A couple of municipal towns (<5%) have taken up disposal of inorganic solid waste by secured landfilling technique.
- Most of the municipalities are practicing uncontrolled open dumping of solid waste.
- Both municipal vehicles and hired vehicles are used for secondary transportation of solid waste.
- Man-power deployment (conservancy workers) ranges between 1 and 3 per 1000 population.
- Average expenditure on SWM in municipal towns ranges between Rs. 32 and Rs. 50 per person per year (average Rs. of 42 per 1000 population).

12.5 Effective Management of Solid Waste

Effective SWM for each town should be developed considering the local situation, scope, available resources, and adaptability. Effective SWM system is needed to safeguard public health and environment. Thus, effective SWM system must be both environmentally and economically sustainable.

The SWM system must ensure 100% collection of solid waste from all sources of generation. Transportation system (both primary and secondary) for SWM should be operated efficiently so that no waste could get accumulated in the town.

The inorganic recyclable solid waste if segregated through institutionalized system will not only fetch money through sale, but also the system will help to uplift socio-economic conditions of scavengers and ragpickers.

There are many options for treatment of solid waste. The organic part of solid waste can be converted to compost which may be used as soil conditioner. Both windrow process of composting and vermicomposting are suitable for the purpose.

The nonrecyclable inorganic solid waste can be disposed in low-yield low-lying land if the disposal system is operated to achieve secured/sanitary landfilling. Land procurement is a challenging task everywhere; as a result, each municipality has to take initiative to procure land for disposal of solid waste for the future. Some municipal towns have already exhausted land for disposal of solid waste, and land acquisition is posing a serious problem to them. In such cases, alternative options, such as pyrolysis/gasification, can be examined to assess feasibility of application. Biochemical conversion of solid waste for production of biogas may be another option which also needs to be examined carefully to assess its field applicability and sustainability.

In order to manage solid waste efficiently, people's participation must be ensured. Involvement of ward committees and schools will be necessary for citizens' awareness and motivation campaign.

12.6 Challenges of Effective Solid Waste Management

One can find certain challenging issues while implementing effective SWM system for the municipal towns. The challenges related to SWM are discussed below.

12.6.1 Segregation of Solid Waste

Since municipal solid waste is heterogeneous in nature, category-wise segregation of the waste at the source of generation is a prerequisite for good management system. Practice of segregation is to be done at household level, and as such social acceptability is an important criterion for its success. To achieve these criteria, awareness, sensitization, and motivation campaigns are required.

12.6.2 Cleanliness of the Town

Maintaining cleanliness of the towns poses a stiff challenge to the city managers. Waste collection from all sources is to be ensured along with abolition of roadside vats. Behavioral change may ensure discipline among the citizens in keeping the city/town clean. Ward communities can take a positive role to keep their wards clean through citizen's active participation.

12.6.3 Waste Minimization

Waste minimization is the process and the policy of reducing the amount of waste produced by the society, institute, and industry. Many ULBs are facing a problem for disposal of solid waste due to land constraint. So, waste minimization or reduction at the source is desirable. To reduce the amount of waste generated at the source, the most practical and promising methods appear to be as follows:

1. The adoption of industry standards for product manufacturing and packaging that uses less material.
2. Enactment of laws that minimize the use of virgin materials in consumer products.
3. The levying (by communities) of cess/fees for waste management services that penalize generators in case of increase in waste quantities.

Modifications in product packaging standard can result in reduction of waste packaging material or the use of recyclable materials. Sorting, recycling, and processing at the source help in waste minimization.

Levying of fees at variable rates for waste collection may promote waste reduction at the source of generation.

12.6.4 Resource Recovery Through Material Recycling

Certain solid waste, e.g., old newspaper, milk and oil pouch, plastic and metal containers, empty cosmetic bottles, broken metal containers, etc., is segregated at household level for selling. Even old garments are exchanged for stainless steel utensils at household level.

Recyclable inorganic solid waste segregation and resource recovery need to be encouraged. The activity must be institutionalized, and scavengers should be socio-economically benefitted. The ULBs must take initiative to institutionalize the whole process. Field estimation highlights that Rs. 22 per year per person can be generated from the sale proceedings of recyclable inorganic waste.

12.6.5 Conversion of Organic Waste to Compost

Biological process facilitates conversion of organic waste to compost by using microorganisms for decomposition of biodegradable components of waste. Windrow composting and vermicomposting are the two appropriate options for converting organic solid waste to compost. The nitrogen, phosphorus, and potassium in compost range between 2.5 and 3.0% which get back to the soil maintaining an ecologically balanced system. The main challenge before ULBs in operating compost plants is the sustainability as farmers are reluctant to purchase compost because of higher cost compared to chemical fertilizers. There must be subsidies in compost production cost to make the process economically viable. The agricultural department should promote the use of compost as good soil conditioner. Sale value of compost manufactured from solid waste generated by an individual (400 gms/c/d) is estimated to be Rs. 73/per year (compost sale value 2000/per MT). But at least 25–30% of subsidy would be necessary on the production cost of compost.

12.6.6 Wastes to Energy

Energy can be recovered from biodegradable as well as non-biodegradable organic fraction of waste through thermochemical and biochemical conversion. The thermochemical conversion processes are useful for waste containing high percentage of organic non-biodegradable matter and low moisture content. The main technology options under this category include incineration and pyrolysis/gasification. Biochemical conversion of organic solid waste could generate biogas as a source of energy.

A decision faced by many communities is determining whether a waste to energy system might be a feasible component of their integrated solid waste management program. It can be lengthy and expensive process for developing a WTE project. It is necessary to follow a step-by-step process for evaluating the feasibility of constructing and operating a WTE facility. The general characteristics of urban solid waste, moisture content, and calorific value indicate that incineration process may not be appropriate for adoption. Feasibility study of other options of WTE needs to be carried out for specific portion of solid waste for assessing their applicability.

12.6.7 Secured Landfilling

Compliance criteria laid down in Municipal Solid Waste (Management and Handling) Rule 2000 indicate that landfilling shall be restricted to non-biodegradable, inert waste, and other wastes that are not sustainable either for recycling or for biological processing. The secured landfill site must be selected, developed, and operated as per guidelines indicated in the above rule. The landfill site should be selected so that it can last for 20–25 years and preferably within 5 km from that present city limits. Pollution prevention is an important challenge during and after completion of landfilling operation. Hence, all preventive measures are necessary to be adopted for landfilling site. Shared disposal of nonrecyclable inorganic and compost plant rejects may be appropriate for a couple of municipal towns if they are situated closely.

12.7 Conclusion

Waste management has become a very complex and challenging task for the urban local bodies. The quantity as well as quality of solid wastes is changing due to technological advancement, consumerism, and behavioral change of the people. The SWM system should be established in a city/town in a planned manner with provision of infrastructural facilities, funding, and peoples' participation. Operational groups responsible for SWM service must be trained and dedicated. Strict discipline in the operation of conservancy services should be ensured for efficient SWM. Solid waste management can never reach the desired level of efficiency until the people participate and discharge their obligation religiously. The ULBs should take up an action plan to modernize SWM and ensure public participation through awareness and motivational efforts. It is the responsibility of the ULBs to establish effective SWM as per Municipal Solid Waste (Management and Handling) Rule 2000 and to obtain authorization from State Pollution Control Board to carry out operational activities.

Prof. Arunabha Majumder was a former director-professor at All India Institute of Hygiene and Public Health, Government of India. Presently, he is an emeritus professor of the School of Water Resources Engineering, Jadavpur University, Kolkata.

Prof. Majumder has more than 40 years of experience in the field of water supply, water purification, water conservation, water quality monitoring, water management, sewerage and drainage, municipal sewage treatment systems, industrial wastewater management, water pollution control, solid waste management, hazardous waste management, environmental impact assessment, etc.

Initially, Prof. Majumder worked at the Calcutta Metropolitan Water and Sanitation Authority and dealt with sewerage and drainage projects, design of sewage treatment plants, solid waste management, etc. Thereafter, he joined All India Institute of Hygiene and Public Health and associated with postgraduate teaching in public health, research and consultancy. Presently, he is emeritus professor at the School of Water Resources Engineering, Jadavpur University.

Prof. Majumder is a fellow of the Indian Water Works Association and Institution of Public Health Engineers (India). He is also a PAC member of the Water Technology Initiative of the Ministry of Science and Technology. He is a member of the Arsenic and Fluoride Task Force, West Bengal.

Dr. Majumder was awarded "Best Environmental Engineer for the Year 2007" by the Institution of Engineers (India). He published more than 95 scientific articles in international and national journals.

Management of Urban Waste Water Infrastructure with Sewer Rehabilitation and Maintenance

13

Nilanghshu Bhusan Basu

13.1 Introduction

Abatement of pollution for preventing deterioration of the environment has become a central issue for quite some time. But the objectives for each problem area towards actual implementation are really diluted due to public indifference so far. The complexities are considerable, given the number of industries, organizations and influx of population. To achieve the objectives and the mix of instruments required in the form of legislations, regulations, physical incentives, voluntary agreement, educational agreement and information campaigns are well known. Though the major emphasis is on increased use of regulation, the focus on alternative technology and applications is really lukewarm. There is an increasing trend in environmental pollution.

Water is one of the worst assaulted assets to receive pollution from traditional economic wastes generated from industrial and domestic processes, chemical agents from fertilizers and pesticides and silt from degraded catchments (Fig. 13.1). It is estimated that 75% by volume of waste water generated is from municipal as well as industrial wastes; it contributes over 50% of the total pollutant, and the major portion of this is coming from large and medium agglomerations. For class 1 cities of this country, less than 5% of the total waste water generation is collected, and less than 25% of this is treated.

Many methodologies were envisaged from time to time to control the waste water in the following manners:

- To prevent pollution at sources
- To encourage, develop and apply the best available practicable technical solutions

N.B. Basu (✉)
Planning & Development Department, The Kolkata Municipal Corporation, Kolkata, India
e-mail: nb.basu.cmc@gmail.com; nb.basu_cme@rediffmail.com

K.J. Nath, V.P. Sharma (eds.), *Water and Sanitation in the New Millennium*,
DOI 10.1007/978-81-322-3745-7_13

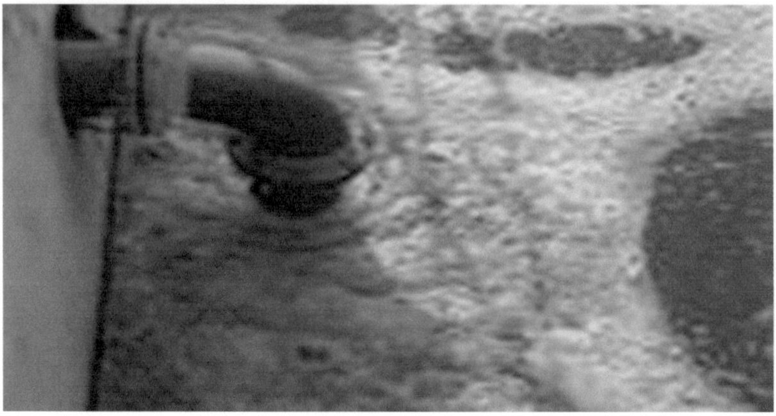

Fig. 13.1 Uncontrolled pollution of water

- To ensure that the polluter pays for the pollution and control arrangements
- To focus protection on heavily polluted areas and river stretches
- To involve the public in decision-making

13.2 Need for the Waste Water Management

In order to handle the waste water, proper management strategy has to be adopted that is suited to the nature and volume of the waste varying from place to place, but in the kind of socio-economic environment, the drainage infrastructure and its maintenance are always having a lower priority. It would not be out of place to mention that urban areas all over the world are having more population concentration than the rural areas around. So the management strategy for waste water in urban areas has to be more stringent than that of rural areas to ensure sanitary condition for healthy environment of the locality (Fig. 13.2). But in reality, it is otherwise in almost all developing countries.

13.3 Identification of Issues

In most of the Indian cities, population concentration is growing at an extremely alarming rate in excess of 20%, so most of the Indian cities are to live with anachronistically low-sized legacy sewers, now old and heavily silted, worn out pumping stations and inefficient and tired inheritance from the same period coupled with unmaintained and uncared for outfall canals that led to encroachments and ruins. Moreover, the maintenance drill starting from the inlet point to the collector sewer, trunk sewer, pumping stations and final outfall is also at a low priority and taking its toll for recurring tragedy of flooding of urban centres every year. The concurrently,

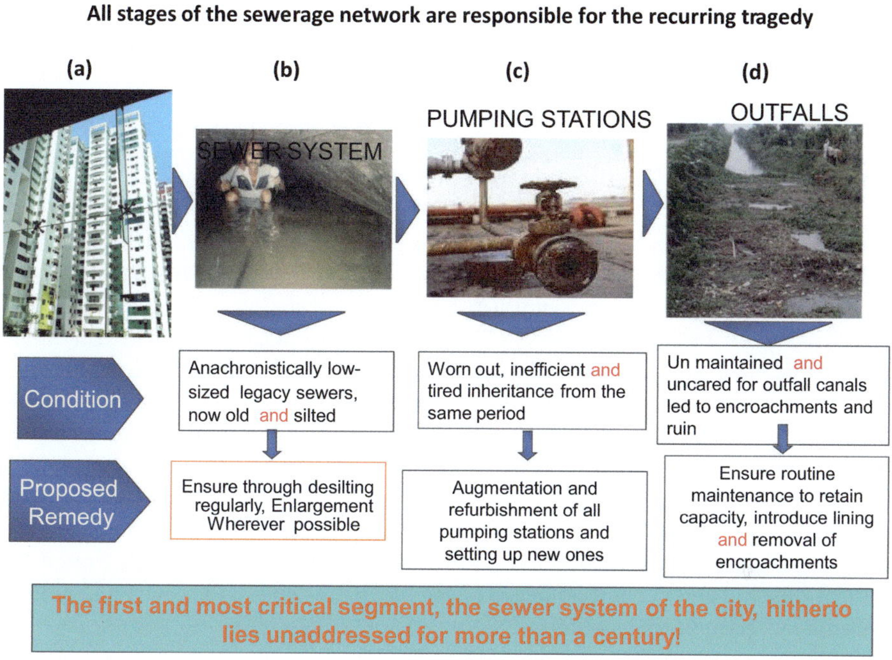

Fig. 13.2 (**a**) Sporadic high rises. (**b**) Silted sewers. (**c**) Worn out old pumps. (**d**) Filled-up outfalls

loss of man-hour, loss of property and health hazard for the city dwellers, cost many folds in terms of loss in GDP, are yet to be recognized.

13.3.1 Major Pollution Sources

Major pollution sources associated with urban drainage and water pollution are the following:

- Industrial and domestic waste water discharged in the river.
- Seventy-five percent of all waste water generated is from municipal sources.
- For class 1 cities of India, less than 5% of total waste water generation is collected, and less than 25% of this is treated.
- This waste water from municipal and industrial establishments, though small in volume, contributes over 50% of the total pollutant loads.
- Lack of public awareness and choked inlet systems.
- The presence of plastic and other non-biodegradable and deleterious substances in the sewer systems.
- An overlay on roads for maintenance is adding to the problem; abutting premises on either side is lowered.

- Failure and poor understanding of the issue.
- Poor state of mitigating measures to upgrade decaying systems from the source to outfall on regular basis.

13.3.2 Collective Results of the Contribution of the Issues

The annual seasonal waterlogging is only a symptom, the malady being of more basic deficiency. What was originally the first link in the house to collector sewer further to trunk sewer for carrying it to the pumping station and finally to outfall chain has now become a source of collapses and catastrophes. Neglect, non-maintenance and anachronistic low size led to solidification of deposits and reduction in hydraulic capacity of the sewers. Silts have become stones, channels are constricted, roofs are collapsing, and joints are falling: city lives on shaky foundation.

The retention of silts for long period of time and indiscriminate dumping of cement slurries and other cementing chemicals down the drains have solidified the deposits of debris making stationary obstructions incapable to be transported by normal flushing and becomes nucleus of big blockades within the sewer and reduce the annular space of the conduits..

13.4 Case Study of Kolkata

City of Kolkata is one of the classical examples, which is blessed with underground drainage of brick sewer system since 1878 and probably the third city in the whole world to have underground drainage system.

Combined sewerage system was commissioned in core Kolkata (the then Calcutta) way back in 1878 with 88 km of man entry sewers and 91 km of nonman entry sewers [1]. The command area of the then Calcutta is like a saucer, depressed at its centre with a natural slope from west to east. So the trunk sewers were laid from west to east, and the same was draining at Bidyadhari River by pumping the same from Palmer Bridge pumping station (PBPS) to finally reach the estuary of Bay of Bengal. Authorities were in a habit of maintaining the gully pits with child labour force and flushing of sewers allowing the flow from the river Hooghly by gravity from the west during high-tide operating gates.

Now the entire system of brick sewers in Kolkata is suffering from severe structural damage and siltation due to years of neglect that has been spelt out earlier.

Urban combined drainage systems installed in older cities across India, especially the ones developed during the colonial era, are provided with combined drainage systems. Since such systems are extremely old and cater to dispose off the combined flow of waste water and storm run-off, they do pose a recipe for significant amount of water pollution at times. It is time that importance of urban drainage be realized by giving due importance in order to safeguard the cities and avoid another catastrophe that took place at Mumbai in July 2005.

Over the time, such practices were gradually withdrawn, and in subsequent decades, desilting of the sewers was done from the manholes and not from the barrel of the conduits. The final outcome was that it was not possible to remove the silts settled along the sewer barrel. As isolation of the sewer sections was not possible being a combined sewer and alternative arrangement for dry weather flow (DWF) from the properties, abutting the roads could not be organized due to narrow and crowded streets. As a result of these, KMC gained an insight into the extremely dilapidated condition of the brick sewers and their high level of siltation, as well as the extent to which these circumstances were contributing to drainage problems and sewer collapse.

The main causes of the deterioration were determined to be the highly limited and localized maintenance work, absence of facilities for screening flows entering sewers, indiscriminate dumping of solid waste and lack of public awareness. The collapse of a sewer is the ultimate stage of its deterioration. Earlier reports indicated that degeneration of bricks and eroded mortar has caused structural distress and failure in several places. Meanwhile, growing traffic loads on overlying roads and excavation of sewers for the purpose of installing other utilities also contributed to sewer collapses.

In fact, sewer collapses increased at an alarming rate between 1980 and 2006 as depicted in (Figs. 13.3, 13.4, 13.5 and 13.6). The cost associated with repairs of such collapses was found to be very high, even not taking into account indirect and social costs.

It would be evident that siltation, collapses and intrusions had frustrated the purpose of the sewers to a great extent, unnoticed by many, who are only interested in the sprawling infrastructure over the ground.

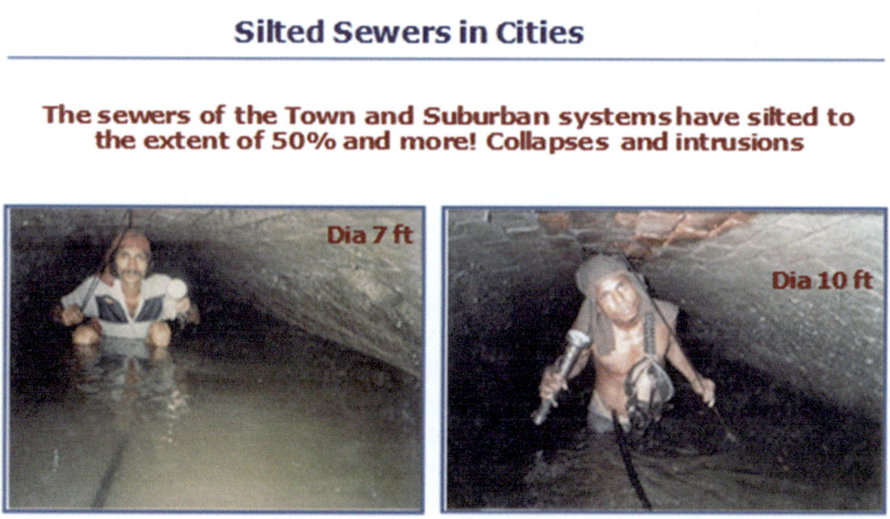

Fig. 13.3 Silted up unattended sewers down underground

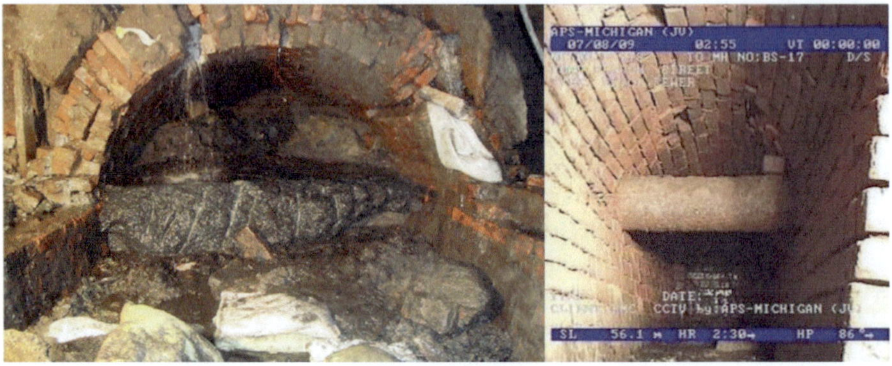

Fig. 13.4 Unattended collapses and intrusions of other services

Fig. 13.5 (**a**) Sewer repair work on a busy street in Kolkata. (**b**) Typical sewer collapse incident

Primary qualities of an efficient drainage system can be summarized as:

- Liquid waste should be disposed off promptly and hygienically.
- Drainage system should be of adequate size and easily cleaned.
- Drainage system should be equipped with liquid seal traps.
- All drains should be adequately ventilated.
- Deleterious substances should be excluded from sewers.
- Backflow of sewage should be prevented.

The poor city managers are caught in the bug of deviations from the classical prescriptions for lack of public knowledge, adverse civic habits, resource

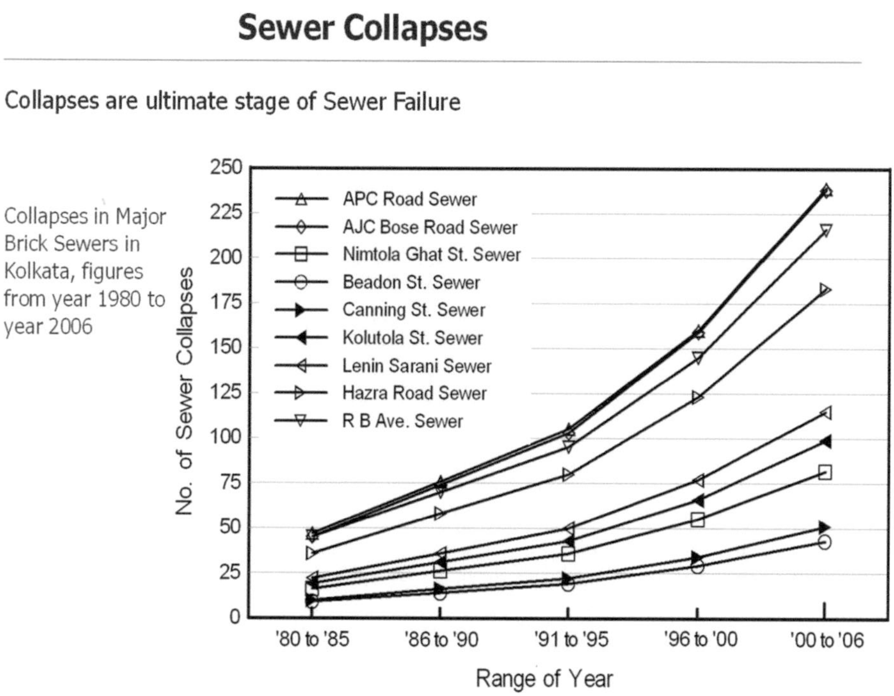

Fig. 13.6 Sewer collapse graph against different year spans

constraints, administrative blockade for selecting state-of-the-art technology and finally lack of political will. All the legacy cities of India are blessed with:

- The annual seasonal waterlogging is only a symptom, the malady being more basic.
- The first link in the house ⇨ sewer ⇨ pumping station ⇨ outfall chain has now collapsed.
- Neglect, nonmaintenance and anachronistic size led to solidification of deposits and reduction in hydraulic capacity of the sewers.
- Silts have become stones, channels are constricted, roofs are collapsing, and joints are falling: city lives on shaky foundation.

In the recent time, all the above difficulties have attracted the attention of the Federal Government of India, and Jawharlal Nehru National Urban Renewal Mission (JNNURM) was launched specially aimed at revamping the urban infrastructure of the legacy cities of India.

13.4.1 Kolkata's Brick Sewers: Critical Issues

Common major issues associated with old sewers in already developed urban centres across India are the following:

- Non-availability of authentic and sufficient data on existing assets.
- Atrociously decaying condition of sewers, both structurally and hydraulically prompting collapses and waterlogging due to heavy siltation.
- Extremely congested urban areas pose logistical challenges and high social cost.
- Certain rehabilitation technologies not yet available in India.
- Lack of public awareness and recognition.

13.4.2 Kolkata's Sewer Rehabilitation Programme

KMC had formulated a challenging project to gradually rehabilitate its brick sewers in trenchless method, and this was earlier approved under JNNURM in 2007. The methodology for the project had to be planned uniquely to suit the field conditions of this densely populated city with overall road coverage of only 6% of land area. Work method followed is as under:

- Topographical, geotechnical and closed-circuit television (CCTV) surveys of sewers were conducted. CCTV surveys were carried out after a sewer had been desilted and again after local repair and then liner installation.
- A gate or barrier was used to isolate a selected stretch of brick sewer. Flows were diverted around the bypassed section of sewer by pumping from the upstream to the downstream manhole through flexible pipes laid on the surface. In addition to standby units and surcharge controls, traffic and other safety precautions including confined space safety were implemented.
- Whenever possible, desilting was conducted by means of a high-pressure jetting machine and a high-volume suction machine. Silt was disposed off at either a municipal dumping ground or a secured sanitary landfill.
- Before sewers were lined, they were repaired, grouted and graded, as necessary, to ensure the required shape and size. The liners were then tested and installed and were later inspected to verify proper installation.
- Manholes were repaired and refurbished, the work site was cleared, and "as-built" drawings were prepared.

13.5 Selection of Technology

Slip-lining (GRP lining) might be the most preferred method of sewer rehabilitation for Indian cities with combined sewer environment. Trenchless method was adopted to create the least possible disruption to urban life, safeguard public safety and manage acute space constraints. CIPP was not suitable for structurally damaged sewers and combined sewers. The two major issues faced in in situ casting of CIPP are:

- Quality control
- Retention of the hydraulic capacity of the sewer to be rehabilitated

Excavation and total sewer replacement is no more an option for its high social cost in densely knit urban areas.

13.5.1 Relevance of Trenchless Technology

Trenchless technology has proved to be a most efficient alternative in densely populated and busy urban roads as the conventional cut and cover method (Fig. 13.7) and not only take too long a time but making the same out of service for all the road users.

The major advantages of trenchless technology are:

- To ensure least disturbance to already existing UG services by using the existing conduit space only for rehabilitation and accessing through access pits at intermittent locations of the existing sewer stretches without digging the entire length (Figs. 13.8 and 13.9)
- To ensure the least social cost

Fig. 13.7 A typical sewer repair by conventional open cut and cover method

Fig. 13.8 Super sucker used for desilting of sewers at night

- To ensure the benefit of connectivity at the quickest time
- To ensure service ability of the existing infrastructure during execution by desilting with super sucker (Fig. 13.8) and structurally rehabilitating by fixing GRP liners in the existing brick conduit (Fig. 13.10)
- To avoid future maintenance of damaged infrastructures for new laying
- To avoid additional cost of relocation of other competing UG services in the working area and at times beyond
- To monitor meticulously

With the extreme challenges negotiated over the past few years, the project has been a success story, as the work of desilting and GRP lining (as per WRC Manual) has been completed and recommissioning of all nine major trunk sewers (Fig. 13.11) taken in the first phase of the programme. These sewers were heavily silted up and had become structurally weak. This initiative has been reported in international engineering journals as front page article (Fig. 13.12). The layout of these nine man entry brick sewers is shown in the map (Fig. 13.13). It is a pleasure to mention that this project, a near perfect example of urban asset management and sustainable infrastructure renewal (Fig. 13.14), has made it to be a featured article of the most circulated civil engineering magazine across the globe [2, 3].

Fig. 13.9 Typical section of access pit for fixing slip lining

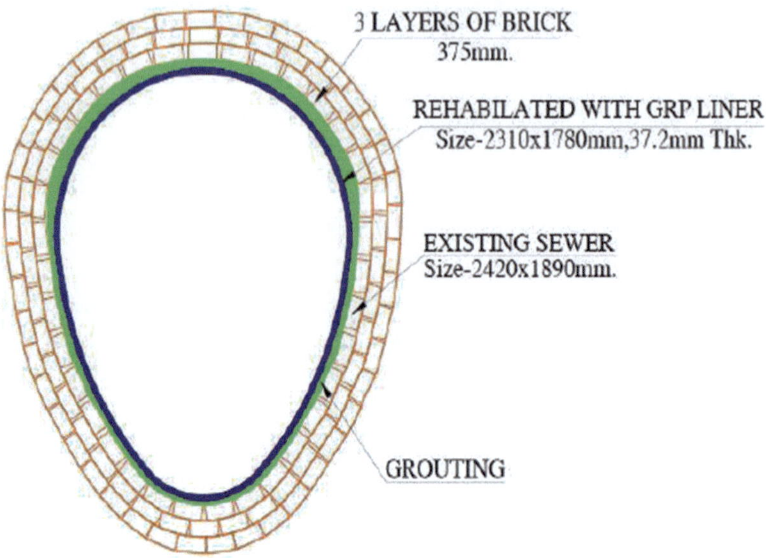

Fig. 13.10 Typical section after rehabilitation with slip lining

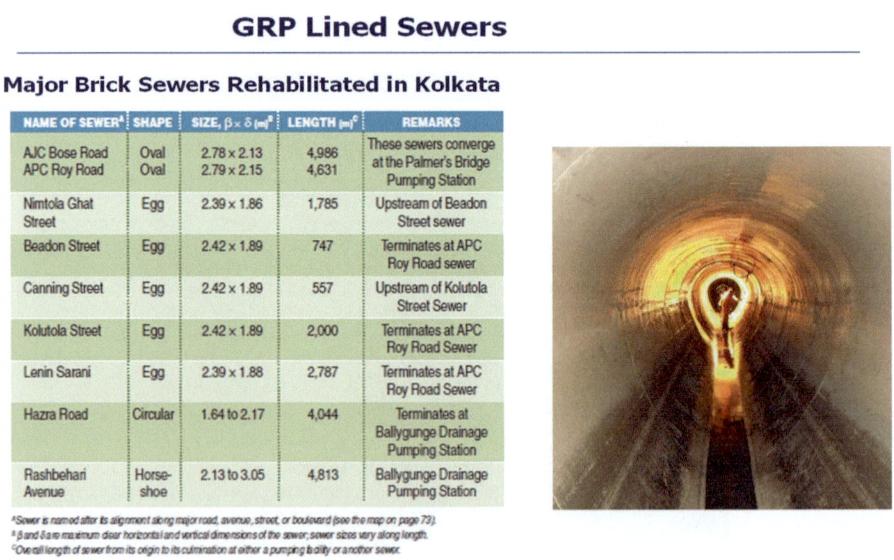

GRP Lined Sewers

Major Brick Sewers Rehabilitated in Kolkata

NAME OF SEWER[A]	SHAPE	SIZE, $\beta \times \delta$ (m)[B]	LENGTH (m)[C]	REMARKS
AJC Bose Road APC Roy Road	Oval Oval	2.78 x 2.13 2.79 x 2.15	4,986 4,631	These sewers converge at the Palmer's Bridge Pumping Station
Nimtola Ghat Street	Egg	2.39 x 1.86	1,785	Upstream of Beadon Street sewer
Beadon Street	Egg	2.42 x 1.89	747	Terminates at APC Roy Road sewer
Canning Street	Egg	2.42 x 1.89	557	Upstream of Kolutola Street Sewer
Kolutola Street	Egg	2.42 x 1.89	2,000	Terminates at APC Roy Road Sewer
Lenin Sarani	Egg	2.39 x 1.88	2,787	Terminates at APC Roy Road Sewer
Hazra Road	Circular	1.64 to 2.17	4,044	Terminates at Ballygunge Drainage Pumping Station
Rashbehari Avenue	Horse-shoe	2.13 to 3.05	4,813	Ballygunge Drainage Pumping Station

[A]Sewer is named after its alignment along major road, avenue, street, or boulevard (see the map on page 73).
[B]β and δ are maximum clear horizontal and vertical dimensions of the sewer; sewer sizes vary along length.
[C]Overall length of sewer from its origin to its culmination at either a pumping facility or another sewer.

Fig. 13.11 Post lining view of major brick sewer in Kolkata

Sewer

Feature article published in American Society of Civil Engineers' (ASCE) *Civil Engineering* magazine (July/August 2012 issue), the most widely circulated engineering magazine in the world reaching > 144,000 civil engineers world wide

New Life for Kolkata's Sewers

Once a pioneer in the development of sewer systems, the Indian city of Kolkata has struggled in recent decades to maintain its collection system network. Rapid population growth and inadequate maintenance led to chronic sanitation and drainage problems, particularly in the city's older urban core. However, Kolkata has embarked on an aggressive program to rehabilitate the oldest sections of its combined sewer system and is already seeing markedly improved system performance and environmental conditions.

By Nilangshu Bhusan Basu and Ayanangshu Dey, Ph.D., CEng, M.ASCE

THE INDIAN CITY OF Kolkata (formerly Calcutta) emerged as a commercial center in the late 17th century, and as it did so it gradually developed an extensive network for collecting wastewater and storm water. Formed in 1876, the Kolkata Municipal Corporation (KMC) assumed responsibility for the city's existing water and sewerage systems, which by the late 19th century included multiple brick sewers of medium to large diameter in what is today the core area of Kolkata. Over time, the sewer network has gradually experienced significant deterioration, the result of age and inadequate cleaning and maintenance. However, in recent years the KMC has begun

Glass-reinforced plastic liners are being installed as part of efforts to rehabilitate a section of Kolkata's extensive network of Victorian-era brick sewers.

Fig. 13.12 A completed section of lined brick sewer

13.6 Conclusion

It is the time to realize the significance of urban drainage systems by giving it due importance in order to safeguard the cities and to avoid further catastrophe that take place each year.

Finally it can be accepted that in an environment of extremely high density of population in urban conglomerations and rapid urbanization in India, there is no option but to improve the civic habits not to throw things around and allow the sewers to function better and longer. The agencies responsible to maintain this absolutely important service would have to be allowed to go for best available technology to reduce the social cost with active participation of all individuals with educated mind.

It would not be out of place to mention that the present method of working is cut-and-cover method. In the present method of working, one has to cut open the road stretch up to a depth of invert level of the drainage conduit in order to have the

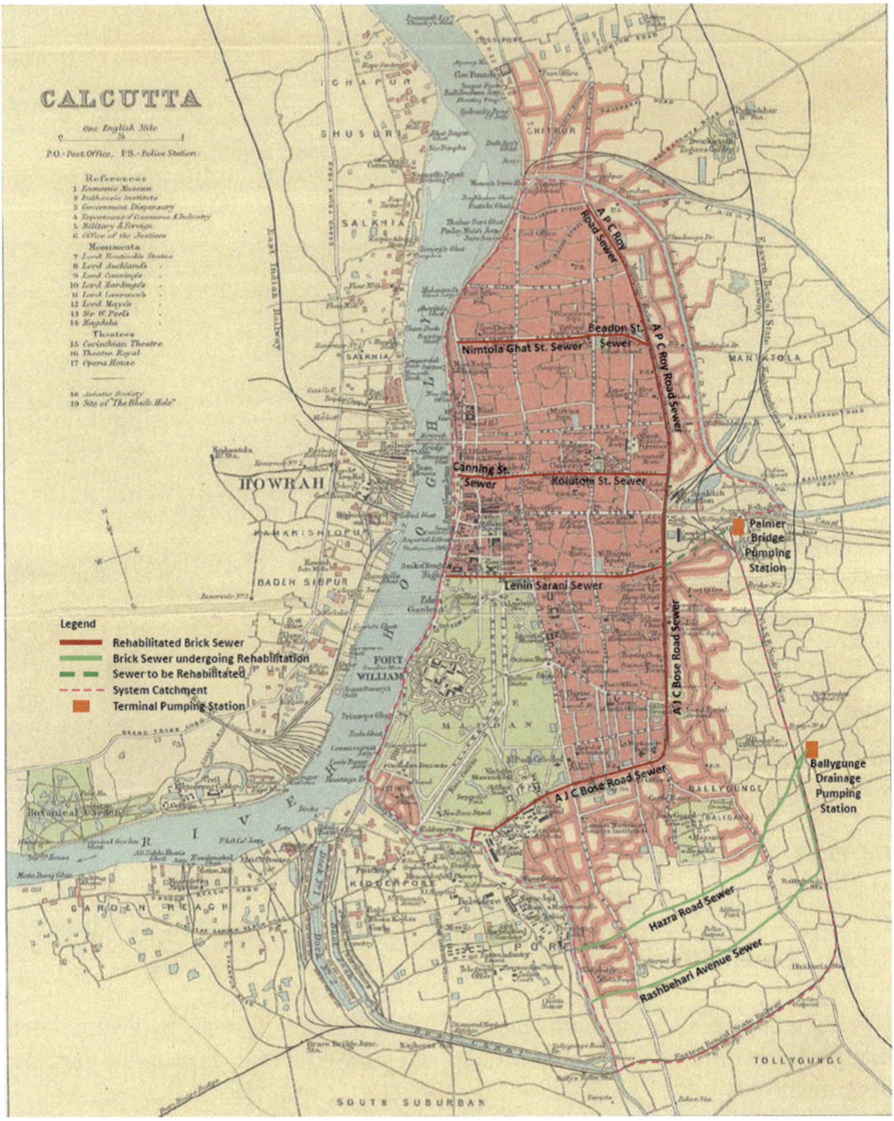

Fig. 13.13 Legacy map of Kolkata showing the brick sewer layout (Photo courtesy: Imperial Gazetteer of India, University of Chicago Library, USA)

foundation of new sidewalls in virgin soil to support the RCC cover slab as the conduit is cleaned opening at its top, and rebuilding of the arch is impossible. In the course of digging and working, many other underground services are disturbed and to be reinstated. The road itself becomes inaccessible for months together at a huge social cost suffering the residents, businesses and visitors of that road on either sides. So finally one would have to resort to trenchless methods, which are yet to popularize in this part of the world in order to become a developed country for the wellbeing of each and all individuals.

Fig. 13.14 Officers inside the rehabilitated sewer

References

1. Goode SW (2005) Municipal Calcutta – Its institutions in their origin and crowth. The Corporation of Calcutta, Macmillan India, New Delhi
2. Basu NB, Dey A (2012) New Life for Kolkata's Sewers. Civil Engineering, American Society of Civil Engineers, July/August 2012
3. Basu NB, Dey A, Ghosh D (2013) Kolkata's brick sewer renewal: history, challenges and benefits. Institution of Civil Engineers, UK, May 2013

Dr. Nilangshu Bhusan Basu Presently working as the Technical Advisor of The Kolkata Municipal Corporation. Former Principal Chief Engineer of The Kolkata Municipal Corporation (KMC) and was the youngest Chief Engineer designate of the state since 1999. I am having a good nos. of engineering papers to my credit in diverse fields of Water Supply, Drainage, Climate Change, Solid Waste Management and Disaster Management. Graduated in Civil Engineering from Bengal Engineering College in the year 1977 with distinction. Completed Post Graduation in Structural Engineering from Jadavpur University in the year 1983 with distinction. Successfully undergone training in River Basin Management at Thames Water Authority, U.K. in the year 1989 with Scholarship. Completed Doctor of Philosophy (PhD) in Environmental sciences from Netaji Subhas Open University in the year 2014. Served as Chairman of Architectural Engg. Division for seven sessions at Institution of Engineers, West Bengal State Centre. Presently adorning the post of Vice President of Institution of Public Health Engineers (I) for last three terms of two years each and again re-elected as Vice President for fourth term.

Role of Fishes in Vector-Borne Disease Control in Southeast Asia

14

Rajpal S. Yadav and Sarfaraz Haq

14.1 Introduction

The high burden of vector-borne diseases such as malaria, dengue, chikungunya and filariasis in Southeast Asia (SEA) and the risk of epidemic diseases require scaling up the coverage of effective interventions and searching for new tools and technologies. Vector control applied through an integrated vector management (IVM) approach can play an important role in controlling vector-borne diseases. The IVM approach promotes the use of ecologically sound and sustainable interventions guided by local evidence and supported through community and inter-sectoral collaboration [1].

At places where mosquito breeding sites can be easily identified, accessed and treated/managed, larval control can be accomplished using chemical or biological larvicides, fishes or larval source management of mosquito aquatic habitats [2]. Such ecological situations include urban areas with high human population density and certain well-demarcated peri-urban/rural settings, such as industrial projects, ports, cantonments, discrete rain-fed borrow pits adjacent to railways and roads and permanent ponds. In semiarid and arid areas where water bodies or man-made reservoirs are limited in number, larval control can be particularly effective.

Larvivorous fishes have been used for mosquito larval vector control for more than 100 years. The application of these fishes to larval breeding habitats in a targeted area can effectively reduce mosquito densities and significantly impact vectorial capacity and disease transmission. In the pre-DDT era, fishes were extensively

R.S. Yadav
Vector Ecology and Management, Department of Control of Neglected Tropical Diseases, World Health Organization, 1211 Geneva 27, Switzerland
e-mail: yadavraj@who.int

S. Haq (✉)
National Institute of Malaria Research (ICMR), Sector 8, Dwarka, Delhi 110077, India
e-mail: sarfaraz_55@rediffmail.com

K.J. Nath, V.P. Sharma (eds.), *Water and Sanitation in the New Millennium*,
DOI 10.1007/978-81-322-3745-7_14

used for controlling mosquitoes [3–5]. Numerous studies have reported the role of fish species such as *Gambusia affinis* and *Poecilia reticulata* and local species such as *Aplocheilus*, *Danio* and *Oryzias* spp. in controlling mosquitoes in a variety of habitats such as wells [6–9], household containers [10], ricefields [11, 12], ponds [13] and wastewater pits [14]. A number of comprehensive reviews have also been published on topics relating to larvivorous fish [15, 16]. The purpose of this article therefore is not to present a review of the literature but to illustrate the role of fish in mosquito control by summarizing outcomes of successful large-scale evaluations, report experiences with operational scaling-up of fishes in national control programme settings and take an overview of the logistic complexities involved in fish-based programmes, with a focus on Southeast Asia.

14.2 Rationale for Larval Control Using Fishes

Knowledge of the ecology of malaria vectors with particular reference to their main breeding habitats is important to understand suitability of larval control by application of fishes. The principal vectors of malaria in Southeast Asia are *Anopheles stephensi*, *An. culicifacies*, *An. fluviatilis*, *An. minimus*, *An. dirus*, *An. annularis*, *An. philippinensis* (= *nivipes*), *An. aconitus*, *An. sinensis* and *An. anthropophagous*. The larval ecology of malaria vectors is summarized in Table 14.1.

The ecological preferences indicate that control of *An. stephensi* by the use of fish is most feasible. It breeds in urban/peri-urban settings and villages having piped water supply in man-made water storage containers and other closed microcosms (wells, covered tanks and buildings), as well as in open bodies of clean water such as concrete reservoirs, man-made recreational pools, garden ponds and industrial sumps. These habitats are identifiable and often accessible for fish application by municipal health services. *An. culicifacies* likes to breed in rural ponds, pools and a variety of small and large open waters in rural and deforested areas. Its control in large bodies is feasible, but fish would not be suitable for use in unaccounted small rain puddles. Breeding of sylvatic vector species such as *An. dirus*, *An. fluviatilis* and *An. minimus* in a wide variety of remote and temporary breeding sites is usually not amenable to larval control. Control of *An. annularis*, *An. philippinensis* (= *nivipes*), *An. aconitus*, *An. sinensis* and *An. anthropophagous* may be technically feasible in specific situations. In non-forested areas with economic activities of fruit tree, rubber and palm plantations in Southeast Asia, *An. dirus* is shown to transmit malaria. The use of fish in water channels and pits with a possible conjugation with biolarvicides to achieve area-wide coverage could be an option to limit breeding of this species.

The larval ecology of dengue and chikungunya vector, *Aedes aegypti*, is very similar to that of *An. stephensi*. The species also breeds in a variety of rainwater collections and small containers that require complementary environmental management. *Ae. albopictus* habitats in rural settings are not suitable for the use of fish. Lymphatic filariasis vector, *Culex quinquefasciatus*, breeds in open wastewater pools, drains, swamps and rice-growing areas. *Mansonioides*, vector of brugian

Table 14.1 The habitats of main malaria vectors in Southeast Asia countries

Species	Country	Eco-epidemiological setting and larval habitats
An. stephensi	India	Urban and rural areas in semi-arid/arid plains; breeds in man-made containers, garden/industrial tanks, recreational reservoirs, tanks at construction sites, pools, wells, rainwater on roof, small, temporary water receptacles/collections in and around buildings
An. culicifacies	India, Sri Lanka	Semi-arid, arid plains with low/moderate endemicity and epidemic potential. Breeds in rural ponds, lakes, wells, canal-seepages, pools at edges of rivers and streams, ricefields, temporary rainwater collections
An. annularis	India, Nepal, Sri Lanka	Coastal, flood-plain and new irrigation areas. Breeds in ponds and pools with aquatic vegetation, rice fields
An. sundaicus	India, Indonesia, Bangladesh, Myanmar, Thailand, Timor Leste	Coastal areas of islands, moderate to high endemicity. Breeds in brackish water in coastal ponds, lagoons
An. fluviatilis	Bhutan, India, Nepal	Forested hills, forest fringes; high endemicity areas. Breeds in streams, river banks, jungle seepages, puddles
An. minimus	NE India, Bhutan, Bangladesh, Myanmar, Thailand, Indonesia	Forest fringes near jungle-cleared areas; high endemicity areas. Breeds in rice fields, edges of streams, running waters, hill seepages
An. dirus	NE India, Bangladesh, Myanmar, Thailand	Deep jungles, across international borders; hyper-endemicity areas. Main habitats are jungle puddles and pools.
An. balabacensis	Indonesia	Perennial malarious deep jungles on islands
An. maculatus	Indonesia, Myanmar, Nepal, Thailand	Peninsular, island or exposed areas with moderate malaria endemicity
An. subpictus	Indonesia, Sri Lanka, Timor Leste	Irrigated, swampy areas.
An. aconitus	Indonesia, Thailand	Rice fields, freshwater fish ponds, streams
An. bancrofti	Indonesia (Irian Jaya)	Coarse reeds, waters with algal growth
An. barbirostris	Timor-Leste	Spring-fed swampy low malaria endemicity areas
An. farauti	Indonesia (Irian Jaya)	Sunlit water collections
An. jeyporiensis	Myanmar	freshwater, slow moving or nearly stagnant with abundant emergent vegetation
An. karwari	Indonesia	Springs, and irrigation canals in rice fields in Irian Jaya
An. koliensis	Indonesia (Irian Jaya)	Grasslands and in pools along the edge of the jungle

(continued)

Table 14.1 (continued)

Species	Country	Eco-epidemiological setting and larval habitats
An. letifer	Indonesia (Kalimantan)	Shaded pools in jungle
An. leucosphyrus	Indonesia (Kalimantan)	Forest pools on island
An. nigerrimus	Indonesia	Canals, large open marshes, large stream pools and rice fields
An. philippinensis	India, Bangladesh, Myanmar	Flood plains in the deltaic regions
An. pseudowillmorei	Thailand	Forest pools
An. punctulatus	Indonesia (Irian Jaya)	Open sunny water collections
An. umbrosus	Indonesia	Kalimantan island, moderate malaria endemicity; swampy forested areas
An. varuna	Sri Lanka	Pools in gem mining areas with high malaria endemicity
An. anthropophagus	DPR Korea	Rice agro-ecosystem in temperate deciduous forest zone
An. sinensis	DPR Korea	Rice agro-ecosystem in temperate deciduous forest zone

filariasis, prefer ponds and pools with aquatic vegetation. *Cx. vishnui*, vectors of Japanese encephalitis, breeds in ricefields, drains and wastewater pools around irrigated systems and in urban/peri-urban areas. In many of these situations, fish may have an independent or a complimentary role in mosquito control.

14.3 Potential Larvivorous Fishes of SEA

A potential larvivorous fish should have a high preference for eating mosquito immatures among a variety of available planktonic food and be small in size, agile and swift, enabling movement in shallow and/or weed-infested waters and escape from predators. Additionally, fish should surface feed, preying upon mosquito larvae that occupy the same strata of the water body, and have high and year-round breeding potential to facilitate the scale-up of fish introduction. Fish should be tolerant to stressful environmental conditions such as elevated temperature, pollutants and water turbidity, capable of withstanding transportation and have little to no commercial value to discourage communities catching them for food.

Based on empirical knowledge, malaria control programmes in Southeast Asia and the neighbouring eastern Mediterranean region have used exotic fish such as *Gambusia affinis*, *Poecilia reticulata* and *Oreochromis* spp. for mosquito vector control. A summary of occurrence of major larvivorous fish species, their qualities, utility in larval control, fitness for mass production and habitats suitable for introduction is given in Table 14.2. Native fishes have also been evaluated for larval control, including those belonging to the genera *Aphanius*, *Aplocheilus*, *Barbus*, *Chela*, *Colisa*, *Danio*, *Esomus*, *Oryzias* and *Rasbora*.

Table 14.2 Occurrence and potential of larvivorous fishes in Southeast Asia region

Genus (no. of species; main species evaluated)	Qualities	Utility in mosquito control	Fitness for mass rearing	Mosquito habitats suitable for release of fish
Aphanius (12 species; *dispar*)	Delicate fishes; measure up to 7.5 cm. in length. Females are smaller than males. Can tolerate high salinity; suitable for both brackish and fresh waters.	Successfully evaluated & utilized at programmatic level in several countries of the Mediterranean region	Can be mass reared in cement tanks/large mud preferably with aquatic vegetation	Open and closed microcosms including brackish waters: ponds, pools, tanks, cisterns, wells
Aplocheilus (6–7 species; *lineatus*)	Small, active, surface feeders. Mainly inhabit fresh waters but also brackish waters of moderate salinity.	Field trials conducted but not yet used at operational level.	No evidence of mass rearing in captivity for mosquito control	Ornamental tanks, fountains, ponds, swamps, wells, reservoirs and overhead tanks
Oryzias (10 species; *melastigma*)	Small sized surface dwelling fishes. Breed yearround in fresh and brackish waters. Widely distributed in eastern coastal states of India and hilly areas with streams	Field trials conducted but not yet used at operational level.	No evidence of mass rearing in captivity for mosquito control. Plenty of fishes available in natural estuarine waters.	Able to control culicines and anophelines. Suitable for use in shallow waters such as rice fields in coastal or hilly areas
Chela (many species; *bacaila*)	Very delicate freshwater fish.	Field trials conducted but not yet used at operational level.	No evidence of mass rearing in captivity for mosquito control	Can be used in ponds, slow running streams and rice fields
Barbus (Puntius) (many species; *ticto*)	Exhibit wide range of size (2.5 to 12.5 cm) and diversified forms. Widely distributed in the region.	Field trials conducted but not yet used at operational level	No evidence of mass rearing in captivity for mosquito control	Suitable for introduction in rain pools, swimming pools, slow streams etc.
Danio (many species)	Surface/sub-surface feeders, abundantly found in foot hills and slow moving streams	Field trials conducted but not yet used at operational level	No evidence of mass rearing in captivity for mosquito control	Suitable for use in shallow freshwater bodies and rice fields
Rasbora (many species; *deniconius*)	Widely distributed in freshwater ponds, hilly streams and tanks with aquatic vegetation	Field trials conducted but not yet used at operational level	No evidence of mass rearing in captivity for mosquito control	Suitable for use in borrow pits, ponds, tanks, fountains

(continued)

Table 14.2 (continued)

Genus (no. of species; main species evaluated)	Qualities	Utility in mosquito control	Fitness for mass rearing	Mosquito habitats suitable for release of fish
Esomus (many species; *denricus*)	Distributed throughout the region in shallow water ponds, tanks, pools, paddy fields and ditches with shallow aquatic vegetations	Field trials conducted but not yet used at operational level.	No evidence of mass rearing in captivity for mosquito control	Suitable for use in marshy areas and rice fields for the control of *Anopheles* and vectors of Japanese encephalitis.
Colisa/Trichogaster (many species; *fasciata*)	A very hardy fish because of the presence of additional air breathing organ. Occur in both fresh and fowl waters, ponds, pools and marshy areas and drains	Field trials conducted but not yet used at operational level.	No evidence of mass rearing in captivity for mosquito control	Suitable for introduction in containers, tanks and industrial effluents
Oreochromis (20 species; *mossambicus*)	Survive in fresh as well as foul waters. Sub-surface feeder	Successfully evaluated and utilized at programmatic level in some African/ Middle-east countries and South-east Asia	Food fhis fish, cultured in ponds in many South-east Asian countries; has been utilised extensivly for mosquito control	Suitable for use in captive ponds, pools, cess-pools, tanks.
Poecilia (many species; *reticulata*)	Viviparous, well known larvivorous fish, now distributed globally	Successfully field tested and utilized at programmatic level in India and many other South-east Asian countries	Can be mass reared in ponds, cement as well as mud tanks.	Can withstand polluted waters; highly suitable for mosquito control in urban drains, factory affluent, wastewater pools, and freshwater habitats
Gambusia (many species; *affinis*)	Viviparous, well known larvivorous fish, now distributed globally. Has shown great potential to reduce mosquito populations throughout the world	Successfully field tested and utilized at programmatic level in India and many other South-east Asian countries.	Can be mass reared in ponds, cement as well as mud tanks.	All type of fresh water mosquito breeding habitats.

14.4 Evidence Base for Use of Fishes

It is often argued that there is lack of sufficient scientific evidence in favour of the use of fishes for vector/disease control. The outcome of some relevant and important field trials to show that in certain ecological situations fish can play an important role either when used alone or as a complementary vector control tool, in reducing vectorial potential or disease transmission, is being reviewed as follows.

14.4.1 Pilot Trials for Evaluation of Efficacy/Effectiveness

14.4.1.1 Control of Mosquitoes and/or Malaria

Results of larvivorous fish trials in eastern Africa (Somalia, Ethiopia, Kenya and Sudan) are described below. A WHO-led cluster-controlled study conducted in the district of Burao, northern Somalia, comprises one of the most scientifically rigorous trials relating to larvivorous fish. The trial was well planned and executed. It demonstrated a high impact on parasitological and entomological indicators. In brief during June 1980–July 1982, the effectiveness of *Oreochromis spilurus spilurus* against malaria was evaluated in 26 villages, spatially located 6–10 km apart [17]. The fishes were collected from local streams 100–208 km away from Hargeisa town. In a cluster of 16 villages (population of 6216), fishes were introduced (1 fish/ m^2) in 608 barkits (man-made reservoirs of rainwater, average surface area of 102 m^2) in June–November 1980. Additionally, fishes were introduced to dry barkits with the beginning of rains during April 1981 and 1982. Five hundred fifty-one barkits in another cluster of ten villages (population of 6024) served as no-fish control. Impact on malaria was assessed by quarterly blood surveys in all villages, reaching 16–28% of the population in treatment villages and 11–18% in control villages. Sentinel villages were included in the study design, with blood films collected from 20–41% persons to 18–44% persons, in treatment and control arms, respectively. All malaria cases (*Plasmodium falciparum*, *P. vivax*, *P. malariae*) were treated. To assess entomological parameters, pyrethrum spray catches of indoor resting *An. arabiensis* and night-biting collections were made. The results of the study indicated an effect of larvivorous fish on both parasitological and entomological indicators. Importantly, the study found that seeding of fishes reduced the pre-intervention malaria parasite rate from 14–20% to 0–1%, as compared with the respective rates of 14–15% and 13–35% in control villages. There were also significant reductions in vector densities and night-biting rates. In subsequent years, there has been great interest in scaling up the use of fishes in barkits and ponds in Somalia, as observed by this author (RSY) during a WHO mission in 2001 (Fig. 14.1).

In Ethiopia, a trial of the indigenous fish, *Aphanius dispar*, was carried out over 12 months (February 1987–January 1988) in the port city of Assab, under the national programme for control of malaria and other vector-borne diseases [18]. Prior to the study, 43–45% of tested wells and household containers (cisterns, barrels, washbasins) contained evidence of mosquito breeding, in particular *An. culicifacies adanensis*. The study found that releasing 5–20 fishes per habitat in 838 sites

Fig. 14.1 A barkit in a Somali village, an ideal mosquito habitat suitable for the use of fish

monthly or biweekly effectively eliminated mosquito larval populations, resulting in 1.6% of habitats found with mosquito larvae as compared with 34% (18–60%) of 776 unstocked sites. Following the success of this trial, the national programme implemented a large-scale larvivorous fish release programme in the city supported by high community participation and collaboration with health and municipal authorities of Assab.

Similar results were found in trials from Kenya and Sudan. In Kenya, release of the larvivorous fish, *Oreochromis niloticus*, in abandoned weed-cleared fishponds produced >94% reduction in larval densities of *An. gambiae* s.l. and *An. funestus* and >75% in culicines over a 15-week observation period [13]. Likewise in Sudan, a 2-year pilot study releasing *Gambusia affinis* in 20 Gezira irrigation canals (1–2 m depth x 4–10 km length) showed marked control of malaria vector *An. arabiensis* as compared with 5 control canals and rapid establishment of fish populations [19]. This effect was mostly seen in dry season, when the canals were also the main source of mosquito production. During the rainy season, mosquitoes breed in rainwater pools not amenable to fish; the study recommended the use of fish in irrigation systems for malaria vector control in the dry season.

A number of studies indicate that the impacts of larvivorous fish on malaria vector control can be broadly generalized beyond the African region. In Grande Comore Island, Comoros, the only breeding habitats of *An. gambiae* s.s. are man-made reservoirs. When *Poecilia reticulata* was introduced in all reservoirs in two island villages, the fish reproduced successfully (obviating the need to restock) and provided yearlong control of larval and adult mosquito populations, significantly reducing malaria incidence in the villages [20]. *P. reticulata* was also introduced in Indonesia, along with the food fish *Cyprinus carpio*. The study, which spanned 25 ricefields and included 112 farmers, lowered the densities of the malaria vector *An. aconitus*, reduced the incidence of malaria and motivated the farmers to coculture larvivorous and food fishes [21].

Some of the most comprehensive operational evaluations of larvivorous fish have been carried out in India from the 1960s to the present day. In Hyderabad city, Sitaraman et al. [22] implemented a large operational release of *G. affinis* and *G. holbrooki*. The fishes were initially introduced in 1966 in 100 wells on an experimental basis and greatly reduced *An. stephensi* and *Cx. quinquefasciatus* larval densities during the 6-month follow-up. Following this, from June 1967–June 1969, the release of the fish gradually replaced the use of Paris green or lead-free petrol for mosquito control in nearly 3800 wells. Chemical larvicides continued to be used in overhead tanks (considered unsuitable for fish) and in isolated cases where consecutive well observations indicated ongoing mosquito breeding. In localities where overhead tanks were regularly treated, there were few cases of malaria, but any disruptions in chemical treatment increased malaria incidence, for example, political disturbances in 1969 disrupted the use of chemical larvicides and malaria rose in the affected localities. In wells, the use of fish continued to reduce breeding of *An. stephensi*, rendering 75% of them free of mosquito larvae. Additionally, the fish survived up to 12 months in the wells and proliferated, contributing to the economic sustainability of the programme. At 1969 prices, the cost of maintaining fish hatcheries in garden and fountain pools was merely INR 15 per month compared with INR 1000 or 300 using petrol or Paris green, respectively. These historical studies have been expanded recently with larvivorous fish field trials in Gujarat, Goa and Karnataka regions in India.

In Gujarat in 1989, a large-scale introduction of guppy fishes in 3815 wells in 100 villages resulted in a drastic reduction in the proportion of wells with larvae (4–7% with larvae) as compared to 889 wells treated with polystyrene beads (16% with larvae) and 117 control wells (23–33% with larvae) [9]. The study demonstrated the high efficacy of fish in controlling the malaria vectors *An. stephensi* and *An. culicifacies* and the filariasis vector *Cx. quinquefasciatus*. More recently, Haq et al. [23] evaluated efficacy of *Aphanius dispar* in controlling mosquito breeding in domestic water storage containers in the command area of a major irrigation project in central Gujarat. A baseline larval survey was done in July 2010 in villages Pithai (n = 100 houses) and Anara (n = 103 houses) and showed the Breteau index of 60 and 69 and container index of 48 and 83%, respectively. In Pithai, fishes were introduced (10–25 fish/tank or container) in 127 underground cement tanks (1 x 1 m to 2 x 3 m, 1.5 m deep) and 167 cisterns (big mud pots and barrels) once during November–December 2010. The village Anara served as control without fish introduction. Over the 12-month study, visual inspection of the containers showed that the fish disappeared in >50% of cisterns due to daily use/replenishment of water by the households, but survived in the large cement tanks (>90% up to 120 days, >70% up to 225 days and >50% up to 360 days of follow-up). There was 63–100% reduction in the density of late-instar larvae of *An. stephensi* and *Ae. aegypti* due to fish application.

In Goa state, two large peri-urban coastal villages (population of 24,252) in PHC Candolim suffered a severe outbreak of malaria in 1993. A study was carried out from 1994 to 1995 to evaluate the vector control potential of *Bacillus thuringiensis israelensis* (*Bti*) and an indigenous fish, *Aplocheilus blockii* [24]. Comparison was made with the incidence of malaria in two adjoining PHCs of Panaji and Porvorim. Pre-intervention larval surveys (January to June 1994) detected the main breeding habitats of malaria vector *An. stephensi* in 1195 wells, 474 overhead tanks, 100 underground tanks, 125 masonry tanks at building construction sites, 107 water sumps, 27 swimming pools, 259 barrels and 4 ornamental fountains, and innumerable curing waters and rainwater pools in construction sites. The residual spraying of DDT indoors and pyrethrum thermal fogging were stopped, and selective application of fish and *Bti* was introduced in July 1994–December 1995. *A. blockii* were applied (5 fish/m^2) in wells, deep masonry tanks, swimming pools and recreational fountains, and a *Bti* H-14 formulation was applied in shallow and temporary habitats not suitable for application of fish such as innumerable temporary rainwater collections especially on terraces, in roof gutters and in curing water on recently cast concrete in buildings and freshly constructed overhead tanks. Malaria surveillance, diagnosis and treatment by the primary health centres were strengthened. The pre-intervention proportion of habitats with mosquito immatures declined from 4.8% (1.2–13.5%) to 0.1–2% during the 18-month period of intervention with fish and Bti. The Bti application in temporary waters reduced proportion of habitats with mosquito immatures from 0.8–3.9% to <0.6%, whereas malaria incidence in PHC Candolim declined by 82% (slide positivity rate, SPR, 17.3% to 7.4%); it continued to increase in the two adjoining PHCs with conventional measures (Panaji, 10–21%; Porvorim, 26.4–27.1%). The study demonstrated the usefulness of combining Bti and fish on a PHC scale in urban malaria control.

In a rural area of Karnataka state, the operational impact of a large-scale introduction of fishes was evaluated from 1994 to 1999 [25]. Fishes were distributed in all suitable water bodies in 93 villages in the district of Kolar. Wells were the producers of the main malaria vector *An. culicifacies* sibling species A, while tanks (ponds) and rain-fed streams mostly contributed to its non-vector siblings and *An. fluviatilis* species T. Malaria was more concentrated in villages with more wells and communities favoured using environmental management and fishes for controlling malaria vector populations in these wells over insecticides due to fears of adverse impacts of insecticides on silkworm rearing and industries. *P. reticulata* were introduced in early 1994 in wells and tanks, but due to their low survival, *Gambusia affinis* were introduced. Entomological monitoring was carried out in two sentinel villages (pre-fish period, 1994–1996; post-fish, 1997–1999). The impact of fish introduction on malaria incidence was monitored fortnightly. Strikingly, malaria incidence was reduced by 99.8% from 1993 to 1999. The larvivorous fish programme is now well established in Karnataka state, and subsequent fish introductions have been guided by the knowledge of the ecology of the sibling species of *An. culicifacies* and *An. fluviatilis* [8], while the use of fishes has also been extended to contain breeding of *Aedes aegypti* in domestic water storage containers [26].

14.4.1.2 Dengue/Chikungunya Vector Control

In Cambodia, in a community-based study from April 2006 to April 2007, *P. reticulata* were introduced in domestic water containers (>200 L household jars and containers) in ~1000 households (14 villages) [27]. In an area 500 m away, 256 households without fish served as control. Twenty-eight village health volunteers were trained who reared guppy fish in jars for distribution to the households in the intervention area. They introduced two to three fishes in each jar and inspected their presence by follow-up visits every fortnight. The presence of fish and larvae was independently assessed by trained entomologists after 1 year. After 1 year of introductions, fishes survived in 57% of containers stocked. An average 10.5% (3.8–22%) of containers had larvae in intervention households as compared to 50% in the control area. Although no baseline *Aedes* indices for the intervention and control areas were reported, there was a 79% reduction in *Aedes aegypti* infestation in the former area as compared to control at the terminal survey. The fishes also caused community-wide effect as the smaller or discarded containers that were unsuitable for fish introduction in the intervention villages had 51% less infestation than those in the control area. In 2008–2009, the larvivorous fish programme was expanded to Bati district, Takeo province, Cambodia, in which a total of 13 health centres and their catchment areas participated [28]. The village health volunteers were trained using the COMBI strategy, health education materials were prepared, and a model for rearing and distributing the guppy fish using community volunteers and local health centres was developed. The communities believed that guppy fishes play an important role in reducing prevalence of *Aedes* mosquitoes and dengue cases.

In China, following an outbreak of dengue in 1980–1981, a community-based programme to control breeding of *Aedes aegypti* in household water containers using indigenous edible Chinese catfish (*Clarias fuscus*) was launched in eight coastal fishing villages in Guangxi Zhuang province [29]. Primary medical personnel supervised the programme. Monitoring of the impact in three sentinel villages showed decline in Breteau index from >47 in 1981 to a very low level in 1985. Contrary to the absence of dengue cases in this area, an outbreak of dengue occurred in the adjoining province of Guangdong in 1985.

In Taiwan, during 1989–1996, integrated control of *Aedes* spp. using fishes in potable water containers, the use of larvicides in vegetable gardens and safe removal of discarded containers from an island of eight villages showed decline of Breteau index from >60 in 1989 to a low level in 1996 [30]. Community motivation was found very effective in this integrated control approach.

In Khon Kaen province, Thailand, the use of fish was found to be an effective method for controlling *Ae. aegypti* in rectangular cement tanks [adjusted odds ratio (AOR), 0.08–0.16] together with correctly covering them with proper lids (AOR, 0.10–0.25) [31]. Keeping fish in containers and correctly covering them with lids were very effective independently, while combining the two increased the effect.

In Karnataka state, India, following an outbreak of *Aedes aegypti*-transmitted chikungunya in 2006, a trial of larvivorous fish was conducted [26]. In two villages in Tumkur district, Karnataka (Domatmari and Srinivaspura), *P. reticulata* was introduced from March to May 2006, while in one village (Balmanda) in Kolar district, *Gambusia affinis* was introduced from July to October 2006 in domestic cement tanks that were the main breeding habitats of *Aedes aegypti*. There was a significant impact on *Aedes* larvae density within 1 week of fish introduction in all the three villages. The clinically diagnosed chikungunya cases were reduced by 65–99% 1 month after fish introduction. The effectiveness of fish was supported by community education. The survival of fish ranged from 16 to 100% after 1 month of introductions in containers.

In Nadiad Taluka of Gujarat state (100 villages), a high proportion of domestic water storage containers were found with breeding of *Aedes aegypti*, *An. stephensi* and *Cx. quinquefasciatus*. *P. reticulata* fishes were introduced in them in 1988 and 1989 on an operational scale for the control of mosquito larvae [32].

14.4.2 Operational Use of Fishes Under Programme Settings

During the pre-DDT era, fishes were used on an operational scale in many countries of eastern Africa, the Middle East [33] and Southeast Asia [16]. They were released in extensive national campaigns such as in Greece for control of malaria vector *An. sacharovi* [4] and have largely contributed to eradication of malaria in Iran [5]. Introduction of *G. affinis* in several identified breeding sites of *Anopheles farauti* complemented with mass drug administration, and treated mosquito nets have eliminated malaria from the island country of Vanuatu [34].

In Central Asia and the Mediterranean region, the main species used are *Tilapia zillii* (Somalia), *Oreochromis mossambicus* (Pakistan), *Carassius auratus* (Iran), *Aphanius dispar* (Oman, Somalia), *Nothobranchius palmqvisti* and *N. guentheri* (Somalia), *Gambusia affinis* (Jordan, Lebanon, Syria, Iran, Iraq, Afghanistan, Sudan) and *Poecilia reticulata* (Pakistan). *A. dispar* has been introduced on an operational scale in Ethiopia and Djibouti and *Oreochromis spilurus spilurus* in northern Somalia for the control of malaria [35, 36].

India has had a vast experience of operational use of fishes for malaria and dengue control. *P. reticulata* and *Gambusia* spp. were imported to India in 1908 and 1928, respectively. They were extensively used in Bombay, Bangalore and Calcutta cities in the control of malaria vector *An. stephensi*. They were abundantly used during the construction of Sharda canal in north India. The success in malaria control due to the use of DDT since the 1950s and later the use of HCH and malathion led to neglect of fish use. It was only around the 1980s when integrated control was revived that the use of fish was picked up in India, particularly when an integrated disease vector control project was initiated in 1983 in Kheda district, Gujarat, that later expanded to other parts of the country.

14.4.2.1 Operational Use of Fish in Kheda Project, Gujarat

A pilot project in Gujarat (1983–1988) evaluated the impact of bioenvironmental control including the use of fish on rural malaria [37]. The project gradually included 261 villages. Interventions included malaria surveillance and case treatment, release of guppy fish or *Gambusia* in permanent water bodies (ponds, wells, canal side pools, ricefields, large cisterns), larval source reduction, sanitation in villages and weeding of pools/ponds. Fishes were mass reared in them often in conjunction with food fishes. A system of fish collection, transportation and restocking in all suitable water bodies was put in place. Expanded polystyrene beads were applied in disused wells. Monitoring of entomological impact was carried all along in sentinel villages. Health education of communities was introduced by village camps. Although the impact of fishes was not separately measured in this integrated approach, entomological evaluations showed a reduction in the breeding potential of the malaria vector *An. culicifacies* and reduction in malaria incidence.

Mass culture of guppies with food fishes (Indian major carps) was undertaken in eight village ponds owned by the village Panchayats. The carps occupy the bottom, middle column and surface of the water body and do not compete for food occupying separate ecological niches. This led to the start of edible fish culture by villagers themselves, and a recent study reported that this activity is being sustained by the communities in the area [38].

Success of the Kheda Project in Gujarat has led to more pilots of this approach elsewhere in India [39]. In the city of Ahmedabad, as part of an operational project on control of malaria and dengue, a model larvivorous fish network was established in 1999–2000 by the municipal corporation in technical support of the National Institute of Malaria Research [40]. Release of fish was guided by a geographical reconnaissance of major mosquito habitats during June 1998–May 1999. Stocks of guppies were raised in the Kankaria lake (10 ha area; Fig. 14.2) and in perennial ponds, garden tanks and abandoned mill tanks. Municipal health staff was trained in fish rearing, collection, transportation and release in water bodies (Fig. 14.3).

Fig. 14.2 Antimalaria staff collecting fishes from Kankaria lake, Ahmedabad

Fig. 14.3 Releasing fish in a fountain pool, a preferred habitat of urban malaria vector *An. stephensi*

Fish distribution tanks were set up at the municipal offices in 43 wards. Community awareness campaigns were organized. Advertisements were given in leading dailies. The people were encouraged to collect fishes from municipal tanks. Guppy fishes were released in all man-made water bodies, viz. cement tanks, ground-level tanks, fountains, wells, mill hydrant tanks, cattle troughs, ponds and various other freshwater pools. There was a sharp decline in the proportion of habitats with larvae at pre-intervention (June 1999, 50–100%) to post-intervention (June 2000, 0–25%) and in malaria cases after the introduction of fishes.

In rural Gujarat, hundreds of fish hatcheries were established, and a training programme on operationalization of fishes was conducted. Following laboratory evaluation of the efficacy of *Aphanius dispar* [41] and its field evaluation in controlling vector mosquitoes in man-made farm ponds (Fig. 14.4) [42], a systematic approach of scaling up the use of the fish was undertaken by the malaria control programme in malaria-endemic Mandvi Taluka in northern Gujarat in November 2008–March 2009 [43]. Over 2.2 million fishes were released in 48 villages in 176 man-made irrigation tanks (khet talavadi), 14 wells, 31 ponds, 9 pools, 13 check dams (small weirs across rain-fed small streams) and 79 other habitats including very large-size pools made from mining excavations in four PHCs areas. Fish introductions resulted in marked reductions in larval densities and malaria incidence in the PHCs.

14.4.2.2 Shahjahanpur, UP

Four hundred *Gambusia* fish hatcheries were established in Shahjahanpur district. Fishes were released in rural wells and ponds reducing larval counts from 60–80%

Fig. 14.4 Farm ponds (khet talavadi) in Kutch, Gujarat, in which *Aphanius dispar* provided long-term and effective control of *An. stephensi and An. culicifacies*

to almost nil. *Gambusia* was successfully cultured along with major carp fishes [44], and the income generated was used for sanitation and development activities in project villages. The operational project showed that composite fish culture can yield benefits of food fish production as well as control of mosquitoes.

The use of larvivorous fishes constitutes an important component of the urban malaria control programme of India. In these areas, vector control operations almost exclusively depend on anti-larval measures, while in certain situations, space spraying is done to contain *Aedes* populations for control of dengue/chikungunya. The national plan for malaria control, 2012–2017, includes biological control using larvivorous fishes as a major element of vector control [45].

14.5 Sustaining Larvivorous Fish Programme

The introduction of larvivorous fish as a vector control intervention has been shown to be economically sustainable in a number of studies. Vector control strategies incorporating larvivorous fish have been integrated with other socio-economic developmental activities across Asia [46]. Examples of this include using aquatic algae for papermaking to generate income and allow fish better access to mosquito larvae and planting eucalyptus to dry marsh areas and provide income resources for vector control, including larvivorous fish stocking. The cocultivation of larvivorous with edible fish in malarious areas has shown promise for reducing larval mosquito populations and generating income for reinvestment into vector control efforts within the community and as an incentive to maintain community involvement in the intervention. A well-documented example of this is seen in Pondicherry, South India, where a sustainable programme for composite fish culture of edible, weed-eating fish and larvivorous fish was initiated by the Vector Control Research Centre to keep perennial rain-fed ponds that irrigate paddy rice free of mosquito larvae and to generate financial gain for the villagers [47, 48]. Similar programmes cultivating larvivorous and edible fish in ricefields in other regions of India [49–51], Indonesia [22] and China [52] have been shown to generate income and significantly reduce malaria incidence. The income generated by such activities can be significant and thus encourage voluntary community participation. In a recent publication, Kant et al. [38] describe programmes carried out from 1985 to 2008 in Gujarat (western India) that emphasized integrated vector management based on the use of larvivorous fish. Income generated from this project exceeded INR 1.25 million over the study period and resulted in the voluntary expansion in the number of villages participating in coculture from 13 to 23 at the time of the report. Larvivorous fishes included guppy, *Gambusia* and *Aplocheilus panchax* and were cocultured with edible fish such as *Labeo rohita*, *Catla catla* and *Cirrhinus mrigala*.

The potential for similar coculture programmes also exists in other locales. Howard et al. [13, 53] note the dual potential of *O. niloticus* as a sustainable source of income and protein to rural farmers and a larval control measure. In surveys of abandoned (fish absent) fishponds, higher mosquito larval densities were found as compared to fishponds containing *O. niloticus* [53]. Field trials reintroducing *O. niloticus* into the abandoned ponds rapidly reduced the number of *Anopheles* and culicine larvae, resulting in >94% reduction in anopheline larvae at 15 weeks following fish introduction. *O. niloticus* fish farming is common in western Kenya and provides both a source of protein and income to local residents. These fishes have low start-up costs and rapid growth, with larger fish able to be harvested for food/income within 6 months of stocking.

Participation of local communities can help sustain the larvivorous fish programmes as was seen in India [40, 54], Djibouti [36], Cambodia [28] and Sri Lanka [55]. A study in Somalia reported good acceptance of the introduction of tilapias (*O. s. spilurus*) as a method of malaria vector control in barkits owned by local households [56]. Success of this study led to introduction of fishes in 20 more villages for malaria control in the region.

14.6 Discussion

The pilot trials and operational experiences with the use of larvivorous fish show that fishes can play an important role in control of the vector-borne diseases. They may be used alone or in conjunction with other measures as part of an integrated vector control strategy. They are specially effective in specific ecological conditions such as in urban, industrial and project areas, in semiarid areas with limited water bodies and in rural areas with water bodies such as ponds, lakes, wells, rain-fed borrow pits, seepage pools along canals, large household domestic water storage containers, irrigation canals and so on.

Benefits of fish use are that they have economic value in income generation and weed control; unlike chemical insecticides, they cannot build resistance in larvae and are safer to use, the communities can be motivated to use them and the operational costs of fish-based programmes are often low. The presence of fish has been shown to prevent gravid *Aedes aegypti* females in laying eggs in containers [57] or prolong developmental times in *Anopheles* larvae [58]. Fish can be an effective control tool due to the low mobility of larval mosquitoes, especially in man-made containers [59].

Survival of the fish is an essential factor to sustain a larvivorous fish programme. The main factors determining the survival of the fish include adaptability of fish to their new environment in which they are introduced. Change in water pH can impact their efficacy [60]. Salinity of water can also influence survival of fish. In Gezira irrigation area, turbid floodwater prevented establishment of *Gambusia* in canals, but they established fairly quickly when released during the dry season when flow of water was modest [19]. Low temperature can hinder establishment and propagation of fish in ponds [61]. Fishes are therefore released in the morning or the evening, water from the new habitat is slowly poured into the container carrying the fishes, fishes are not released at the time of flooding in the rainy season and carnivorous fishes are netted out from village ponds. Aquatic vegetation and weed that can hinder fish movement and provide refuge for the mosquito larvae may be removed manually [13] or using weed-eating fishes to aid in fish establishment [62]. Clearance of aquatic vegetation enables fish to access mosquito larvae and enhance the impact on mosquito populations.

A ditch-ridge system for ricefields to accommodate the fish was recommended by Wu et al. [52] Special considerations must be given to fish survival in water storage containers and in fields. Excessive chlorination of piped water stored in containers [56] or the use of pesticides and fertilizers in irrigated fields can adversely affect fish survival [63]. The restocking of fish in water storage containers may be required in urban areas due to chance of physically removing fishes during cleaning [18] or mortality caused by chlorinated water or high temperature. Community education about the benefits of fish is essential to ensure propagation or survival of fish in stocked sites.

In certain situations, where fishes are not suitable for use, application of *Bti* [24, 36] or chemical larvicides [22] has been shown to complement efficacy of fishes. Control of pathogens is required to prevent infections in fishes in hatchery ponds [33]. For long-distance transportation, fishes are packed in plastic bags with oxygen [38]. If they are carried in open containers by road in tropical climates, speed of the carrier vehicle should be kept modest to limit turbulence in water and vehicles frequently stopped under the shade.

The local species being well adapted to their environment can be equally efficacious as the exotic species [41, 64]. Effectiveness of native fish species in controlling vector-borne diseases should further be explored. A successful fish-based programme requires capacity strengthening for larval reconnaissance and knowledge of larval ecology, mapping and rationale targeting of habitats and human and financial resources to ensure continuation of institutional support to the communities. Malaria programmes should invest in development of training resources for fish collection/mass rearing, their transportation and release, monitoring of fish survival and impact and when required construction of hatcheries and their maintenance. Community participation and inter-sectoral collaboration such as with rural development programmes, irrigation departments, development projects, municipalities, fisheries and universities/research institutions are essential in the success of a fish-based larval control programme using fish in an IVM context. Applied and operational research is also needed to evaluate the effectiveness of potential larvivorous fish. Furthermore, scientifically and statistically valid study designs should be used in any future evaluations of the impact of larvivorous fishes.

Acknowledgement The views and opinion expressed in this paper are solely those of the authors and do not necessarily represent policies or views of their respective organizations. The authors declare no conflict of interest.

References

1. WHO (2012) Handbook for integrated vector management. World Health Organization, Geneva
2. WHO (2013) Larval source management: a supplementary measure for malaria vector control. An operational manual. World Health Organization, Geneva
3. Floore TG (2006) Mosquito larval control practices: past and present. J Am Mosq Control Assoc 22:527–533
4. Hadjinicolaou J, Betzios B (1973) *Gambusia* fish as a means of biological control of *Anopheles sacharovi* in Greece. WHO unpublished document WHO/VBC/73. 463: 1–7
5. Tabibzadeh I, Behbehani G, Nakhai R (1970) Use of *Gambusia* fish in the malaria eradication programme of Iran. Bull WHO 43:623–626
6. Sharma SK, Rajagopai R (1986) Efficacy of fish larvivore, *Gambusia affinis* for the control of mosquito breeding in wells in semi-urban areas near Delhi. J Com Dis 18:95–102
7. Menon PKB, Rajagopalan PK (1978) Control of mosquito breeding in wells by using *Gambusia affinis* and *Aplocheilus blockii* in Pondicherry town. Indian J Med Res 68:927–933
8. Ghosh SK, Tiwari SN, Sathyanarayan TS, Sarnpath TRR, Sharma VP, Nanda N et al (2005) Larvivorous fish in wells target the malaria vector sibling species of the *Anopheles culicifacies* complex in villages in Karnataka, India. Trans R Soc Trop Med Hyg 99:101–105
9. Rajnikant, Gupta DK, Sharma HC, Srivastava HC, Gautam AS (1993) Observations on mosquito breeding in wells and its control. Indian J Malariol 30(4):215–220
10. Neng W, Shusen W, Guangxin H, Rongman X, Guangkun T, Chen Q (1987) Control of *Aedes aegypti* larvae in household water containers by Chinese cat fish. Bull WHO 65:503–506
11. Lacey LA, Lacey CM (1990) The medical importance of riceland mosquitoes and their control using alternatives to chemical insecticides. J Am Mosq Control Assoc Suppl 2:1–93
12. Yadav RS, Das MK (1994) Role of *Danio* and *Oryzias* fishes in the control of mosquito breeding in rice fields. In: Sharma VP, Ghosh A (eds) Larvivorous fishes of inland ecosystems. Malaria Research Centre, Delhi, pp 61–68
13. Howard AFV, Zhou G, Omlin FX (2007) Malaria mosquito control using edible fish in western Kenya: preliminary findings of a controlled study. BMC Public Health 7:199. doi:10.1186/1471-2458-7-199

14. Chand SK, Yadav RS (1994) Use of *Oreochromis mossambicus* (Peters) in controlling mosquito breeding in cowdung pits. In: Sharma VP, Ghosh A (eds) Larvivorous fishes of inland ecosystems. Malaria Research Centre, Delhi, pp 115–120
15. Gerberich JB, Laird M (1968) A bibliography of papers relating to the control of mosquitoes by the use of fish. An annotated bibliography for the years 1901-1966. FAO Fish Tech Pap 75:70
16. Sharma VP, Ghosh A (1994) Larvivorous fishes of inland ecosystem. Malaria Research Centre (ICMR), Delhi
17. Alio AY, Isaq A, Delfini LF (1985) Field trial on the impact of *Oreochromis spilurus spilurus* on malaria transmission in northern Somalia. Unpublished documents, World Health Organization, WHO/MAL/85.1017 and WHO/VBC/85.910
18. Fletcher M, Teklehaimanot A, Yemane G (1992) Control of mosquito larvae in the port city of Assab by an indigenous larvivorous fish, *Aphanius dispar*. Acta Trop 52:155–166
19. Mahmoud AA (1985) Mosquitofish *Gambusia affinis* Holbrooki as a malaria vector control agent in Gezira irrigation canals of the Sudan. J Am Mosq Control Assoc 1(4):524–526
20. Sabatinelli G, Blanchy S, Majori G, Papakay M (1991) Impact de l'utilisations du poisson larvivore Poecilia reticulata sur la transmission du paludisme en RFI des Comores. Ann Parasitol Hum Comp 66:84–88. [in French]
21. Nalim S, Boewono DT, Haliman E (1988) Control demonstration of the rice field breeding mosquito *Anopheles aconitus* Donitz in Central Java, using *Poecilia reticulata* through community participation: 3. Field trial and evaluation. Bul Penelit Kesehat 16(1):6–11
22. Sitaraman NL, Karim MA, Reddy GV (1975) Observations on the use of *Gambusia affinis* Holbrooki to control *A. stephensi* breeding in wells. Results of two years' study in Greater Hyderabad city–India. Indian J Med Res 63:1509–1516
23. Haq S, Srivastava HC (2013) Efficacy of *Aphanius dispar* (Rüppell) an indigenous larvivorous fish for vector control in domestic tanks under the Sardar Sarovar Narmada project command area in District Kheda, Gujarat. J Vector Borne Dis 50(2):137–140
24. Kumar A, Sharma VP, Sumodan PK, Thavaselvam D (1998) Field trials of biolarvicides *Bacillus thuringiensis israelensis* strain 164 and the larvivorous fish *Aplocheilus blocki* against *Anopheles stephensi* for malaria control in Goa. Am J Mosq Cont Assoc 14(4):457–462
25. Ghosh SK, Tiwari SN, Kulshreshtha AK, Sathyanarayan TS, Sampath TRR (2002) Control of malaria transmission using larvivorous fishes. In: Raghunath D, Nayak R (eds) Trends in malaria and vaccine research: the current Indian scenario. Tata McGraw-Hill Publishing Company Ltd, New Delhi, pp 154–158
26. Ghosh SK, Chakaravarthy P, Panch SR, Krishnappa P, Tiwari S, Ojha VP et al (2011) Comparative efficacy of two poeciliid fish in indoor cement tanks against chikungunya vector *Aedes aegypti* in villages in Karnataka, India. BMC Public Health 11:599
27. Seng CM, Setha T, Nealon J, Socheat D, Chantha N, Nathan MB (2008) Community based use of the larvivorous fish *Poecilia reticulata* to control the dengue vector *Aedes aegypti* in domestic water storage containers in rural Cambodia. J Vector Eco 33:139–144
28. ADB/WHO (2013) Managing regional public goods for health. Community-based dengue vector control. Asian Development Bank and World Health Organization, Manila
29. Wu N, Wang S, Han G, Xu R, Tang G, Qian C (1987) Control of *Aedes aegypti* larvae in household water containers by Chinese catfish. Bull World Health Organ 65:503–506
30. Wang CH, Chang NT, Wu HH, Ho CM (2000) Integrated control of the dengue vector *Aedes aegypti* in Liu-Chiu village, Ping-Tung County, Taiwan. J Am Mosq Control Assoc 16:93–99
31. Phuanukoonnon S, Mueller I, Bryan JH (2005) Effectiveness of dengue control practices in household water containers in Northeast Thailand. Tropical Med Int Health 10:755–763
32. Gupta DK, Bhatt RM, Sharma RC, Gautam AS, Rajnikant (1992) Intradomestic mosquito breeding sources and their management. Indian J Malariol 29(1):41–46
33. WHO (2003) Use of fish for mosquito control. Document WHO-EM/MAL/289/E/G. WHO Regional Office for the Eastern Mediterranean, Cairo
34. Kaneko A, Taleo G, Kalkoa M, Yamar S, Kobayakawa T, Björkman A (2000) Malaria eradication on islands. Lancet 356:1560–1564

35. Fletcher M, Teklehaimanot A, Yemane G, Kassahun A, Dane GK, Beyene Y (1993) Prospects for the use of larvivorous fish for malaria control in Ethiopia: search for indigenous species and evaluation of their feeding capacity for mosquito larvae. J Trop Med Hyg 96(1):12–21
36. Louis JP, Albert JP (1988) Malaria in the Republic of Djibouti. Strategy for control using a biological antilarval campaign: indigenous larvivorous fish (*Aphanius dispar*) and bacterial toxins. Med Trop (Mars) 48(2):127–131. [In French]
37. Sharma VP, Sharma RC, Gautam AS (1986) Bio-environmental control of malaria in Nadiad, Kheda District, Gujarat. Indian J Malariol 23:95–117
38. Kant R, Haq S, Srivastava HC, Sharma VP (2013) Review of the bioenvironmental methods for malaria control with special reference to the use of larvivorous fishes and composite fish culture in central Gujarat, India. J Vector Borne Dis 50(1):1–12
39. NIMR (2007) Integrated disease vector control project. A profile. National Institute of Malaria Research, Delhi
40. Haq S, Yadav RS, Kohli VK (2003) Developing larvivorous fish network for mosquito control in urban areas: a case study. ICMR Bull 33(7):69–73
41. Haq S, Yadav RS (2011) Geographical distribution and evaluation of mosquito larvivorous potential of *Aphanius dispar* (Rüppell), a native fish of Gujarat, India. J Vector Borne Dis 48(4):236–241
42. Yadav RS (2006) Community randomized trial for malaria control using a native fish–*Aphanius dispar*. NIMR Newslet 1:5–6
43. Haq S (2013) Operational scaling-up use of larvivorous fishes and its impact on prevalence of malaria in Mandvi Taluka of District Kutch, Gujarat. J Trop Dis 1:108. doi:10.4172/jtd.1000108
44. Haq S, Prasad RN (1991) Culture of *Gambusia affinis* with food fishes. Indian J Malariol 28:201–206
45. NVBDCP. Strategic Plan for Malaria Control in India 2012–2017. A Five-year Strategic Plan. Directorate of National Vector Borne Disease Control Programme, Delhi
46. Manderson L, Valencia L, Thomas B (1992) Community participation and the control of tropical diseases. Document TDR/SER/RP/92/1, TDR, Geneva
47. Rajagopalan PK, Panicker KN (1986) Vector control: how to gain acceptance and support from the community. WHO Chron 40(5):184–187
48. Rajagopalan PK, Panicker KN, Das PK (1987) Control of malaria and filariasis vectors in South India. Parasitol Today 3(8):233–241
49. Victor TJ, Chandrasekaran B, Reuben R (1994) Composite fish culture for mosquito control in rice fields in southern India. Southeast Asian J Trop Med Pub Health 25(3):522–527
50. Gupta DK, Sharma RC, Sharma VP (1989) Bioenvironmental control of malaria linked with edible fish production in Gujarat. Indian J Malariol 26:55–59
51. Haq S, Yadav RS, Dash AP (2015) Composite fish culture with larvivorous fishes and environmental interventions for vector control. In: Sharma VP, Vasudevan P, Satya S, Sharma V (eds), Women and health care: initiatives in vector control through community mobilization. Indian Institute of Technology, New Delhi, pp 112–120
52. Wu N, Liao G, Li D, Luo Y, Zhong G (1991) The advantages of mosquito biocontrol by stocking edible fish in rice paddies. Southeast Asian J Trop Med Pub Health 22:436–442
53. Howard AFV, Omlin FX (2008) Abandoning small-scale fish farming in western Kenya leads to higher malaria vector abundance. Acta Trop 105(1):67–73
54. Sharma VP, Sharma RC (1989) Community based bioenvironmental control of malaria in Kheda district, Gujarat, India. J Am Mosq Control Assoc 5:514–521
55. Van den Berg H, Hildebrand AV, Ragunathan V, Das PK (2007) Reducing vector-borne disease by empowering farmers in integrated vector management. Bull World Health Organ 85:561–566
56. Mohamed AA (2003) Study of larvivorous fish for malaria vector control in Somalia, 2002. Eastern Med Health J 9(4):618–626
57. Pamplona LGC, Alencar CH, Lima JWO, Heukelbach J (2009) Reduced oviposition of *Aedes aegypti* gravid females in domestic containers with predatory fish. Tropical Med Int Health 14:1347–1350

58. Bond JG, Arredondo-Jimenez JI, Rodriguez MH, Quiroz-Martinez H, Williams T (2005) Oviposition habitat selection for a predator refuge and food source in a mosquito. Ecol Entomol 30:255–263
59. Killeen GF, Fillinger U, Knols BGJ (2002) Advantages of larval control for African malaria vectors: low mobility and behavioural responsiveness of immature mosquito stages allow high effective coverage. Malar J 1(8):1–7
60. Ahmed TU, Bhuiyan KR, Meher-E-Khoda (1985) Observations on the larvivorous efficiency of *Poecilia reticulata* rosen and bailey (cyprinodontiformes: cyprinidontidae). Bangladesh J Zool 13(1):7–12
61. Joshi GC, Wattal BL, Bhatnagar (1978) Efficacy of larvivorous fish *Poecilia reticulatus* (Peters) for the control of *Culex pipiens fatigans* Wied in a rural area of Delhi. J Com Dis 10(4):238–243
62. Jayasree M, Sabesan S, Panicker KN (1990) Weedivorous fishes for the control of vectors of Malayan filariasis. Indian J Med Res 91:379–381
63. Lacey LA, Lacey CM (1990) The medical importance or riceland mosquitoes and their control using alternatives to chemical insecticides. J Am Mosq Control Assoc 6(Suppl):1–93
64. Nelson SM, Keenan LC (1992) Use of an indigenous fish species, *Fundulus zebrinus*, in a mosquito abatement program: a field comparison with the mosquito fish, *Gambusia affinis*. J Am Mosq Control Assoc 8:301–304

Rajpal S. Yadav is a medical entomologist with a more than 30 years of laboratory and field experience. He was formerly head of the field station of the National Institute of Malaria Research, Gujarat, India. His work in India and South-East Asia involves developing innovative technology of integrated vector control, epidemiology of malaria and dengue, insecticide resistance monitoring, capacity strengthening in vector control, epidemic containments and developing policies and strategies for malaria and dengue control. He currently heads the WHO Pesticide Evaluation Scheme at WHO HQ, Geneva, Switzerland where his work involves evaluation of new vector control tools, developing global policies, strategies and guidelines relating to vector control and sound management of pesticides, and country support in vector control. Dr. Yadav has been a Fellow of the Indian Society for Malaria and other Communicable Diseases (FISCD) and Fellow of the Royal Entomology Society. He has authored over 85 research papers and led development of several WHO guidelines on evaluation of vector control products.

Sarfaraz Haq is fish biologist who worked for nearly 30 years at the National Institute of Malaria Research (field stations Shahjahanpur, UP, and Nadiad, Gujarat). He has conducted extensive fish fauna surveys to identify indigenous larvivorous fishes in India and is credited with developing operational systems for scaling up use of fishes for vector control in collaboration with national vector-borne disease control programme in India. He has published several papers on larvivorous fishes.

Vector Control: Need for an Integrated and Eco-based Approach

15

Vinod Prakash Sharma

15.1 Introduction

In India malaria, lymphatic filariasis, dengue and dengue haemorrhagic fever, chikungunya, Japanese encephalitis and kala-azar are common vector-borne diseases. The nationwide control of these diseases is organized by the National Vector-Borne Disease Control Programme (NVBDCP) of the Ministry of Health and Family Welfare. Since health is a state subject, central government provides technical guidance and shares financial burden; the state governments are responsible for field operations. NVBDCP was initially responsible for malaria control with some support to other vector-borne diseases, but since last decade, all vector-borne diseases have been brought under the NVBDCP. The logic behind this merger was that vector control techniques do not change, but their application in the field may require some modification depending on the vectors involved. Therefore the system of integrating all vector-borne diseases under one umbrella would reduce cost of field operations and provide experienced technical staff for planning and control of these diseases. It may be noted that there are several studies that contradict NVBDCP's morbidity and mortality figures highlighting gross under-reporting. However, the available NVBDCP's data would provide adequate information for planning and execution of the integrated eco-based approach in the management of vector-borne diseases in India. The latest information uploaded by the NVBDCP in respect of the vector-borne diseases' incidence is given below (NVBDCP website).

Malaria Malaria cases in 2010 were 1,599,986 (Pf 834,364) and deaths 1018. In 2009 the total malaria cases reported by 19 states under the UMS were 166,065 (Pf 31,134) and deaths 213. There are six major and two minor malaria vectors, viz.

V.P. Sharma (deceased)
Centre for Rural Development and Technology, Indian Institute of Technology, Hauz Khas, New Delhi 110016, India
e-mail: vinodsharma1938@gmail.com

K.J. Nath, V.P. Sharma (eds.), *Water and Sanitation in the New Millennium*,
DOI 10.1007/978-81-322-3745-7_15

Anopheles culicifacies, *An. stephensi*, *An. fluviatilis*, *An. minimus*, *An. baimaii* and *An. sundaicus*; *An. annularis* and *An. philippinensis* are secondary vectors. All vectors breed in clean water either stagnant or slow-flowing or brackish water.

Lymphatic Filariasis 250 endemic districts in 20 States/UTs. Endemic population 600 million, of this 21 million with symptomatic filariasis and 27 million with microfilaria carriers. The vector *Cx. quinquefasciatus* breeds in polluted waters, sewage, silage, surface drains, cuss pools, cuss pits, drains and septic tanks. *Cx. quinquefasciatus* can bred in relatively clean water. Main nuisance mosquito throughout rural and urban India.

Dengue and Dengue Haemorrhagic Fever In 2010, 28,292 cases and 108 deaths. Vectors *Aedes aegypti* and *Aedes albopictus* breed in desert coolers, dams, jars, pots, flower vases, containers of any size, tanks, cisterns, bottles, solid wastes, tyres, roof gutters, cemetery urns, bamboo stumps, coconut shells, tree holes and many more places where rainwater collects.

Chikungunya In 2010, 48,176 cases. Breeds in conditions given in case of dengue. Vector *Aedes aegypti* and *Aedes albopictus*.

Japanese Encephalitis First JE case was reported in 1955. Outbreaks have been reported from different parts of India. From 1998 to 2004, 15 states and union territories have reported JE incidence. Annual JE incidence ranged between 1714 and 6594 and deaths between 367 and 1665.

Visceral Leishmaniasis or Kala-Azar In 2010, 29,000 cases and 105 deaths. Sandfly of genus *Phlebotomus argentipes* is the only known vector of kala-azar in India. Sandflies prefer high relative humidity, warm temperature, high subsoil water and abundance of vegetation and breed in places with high organic matter that serves as food for larvae.

This paper is primarily based on the practical experience on malaria and its control in India. Malaria is a major public health problem, and almost 90% population is at risk of contracting malaria. Although currently insecticides are the mainstay of malaria control, but for sustainable and cost-effective malaria control, integrated vector management (IVM) is more appropriate and desirable.

15.2 Malaria Vectors

In India there are six major and two minor malaria vectors. Figure 15.1 shows the distribution of major malaria vectors in India. These are the *Anopheles culicifacies*, *An. stephensi*, *An. fluviatilis*, *An. minimus*, *An. baimaii* and *An. sundaicus*. Minor vectors are *An. annularis* and *An. philippinensis*. All above vectors transmit *Plasmodium vivax*, *Plasmodium falciparum* and *Plasmodium malariae*; *P. ovale* and *P. knowlesi* (recently added in the list of human malaria parasites) do not occur

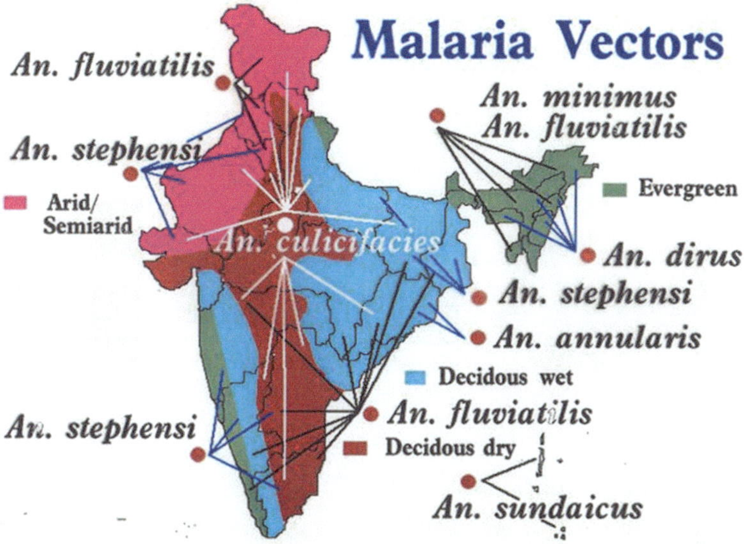

Fig. 15.1 Distribution of malaria vectors in India (Source: NIMR, New Delhi)

in India. Malaria vectors have geographic distribution specific to each species, as shown in Fig. 15.1. Malaria control in India is primarily based on the spraying of residual insecticides in rural areas and source reduction and larviciding in urban areas. Several insecticides have been used in the control of rural malaria, viz. DDT, HCH, malathon and synthetic pyrethroids, etc. The main problem is the development of insecticide resistance in malaria vectors and use of replacement insecticides which eventually leads to double or multiple insecticide resistance [1, 2].

Anopheles culicifacies is the vector of rural malaria in the entire plains of India. In areas with slow-flowing streams, *An. fluviatilis* also breeds and transmits malaria but found mostly near the foothills. In many situations, e.g. in UP Terai, malaria transmission begins with the rainy season when almost all malaria transmission is by *An. culicifacies*, but as the summer season cools, the breeding shifts to *An. fluviatilis* and terminates with the onset of winters. The two vectors maintain almost a relay transmission [3]. In-depth studies on the biology of these vectors revealed that *An. culicifacies* is a complex of five sibling species designated as A, B, C, D and E (species D is also known by the name of *An. dirus*). Figure 15.2 gives the distribution of *An. culicifacies* sibling species in India. Sibling species are morphologically indistinguishable mosquitoes with reproductive isolation. One or more sibling species can coexist in nature in complete isolation without producing F1 progeny. The distribution of species and their sibling species regulates malaria transmission. This is due to feeding and sporogonic development that is influenced by the environment. For example, *An. culicifacies* has five sibling species designated as A, B, C, D and E. Species B is either a nonvector or extremely poor malaria vector [4]. All other sibling species are vectors, and their potential to transmit malaria varies from one

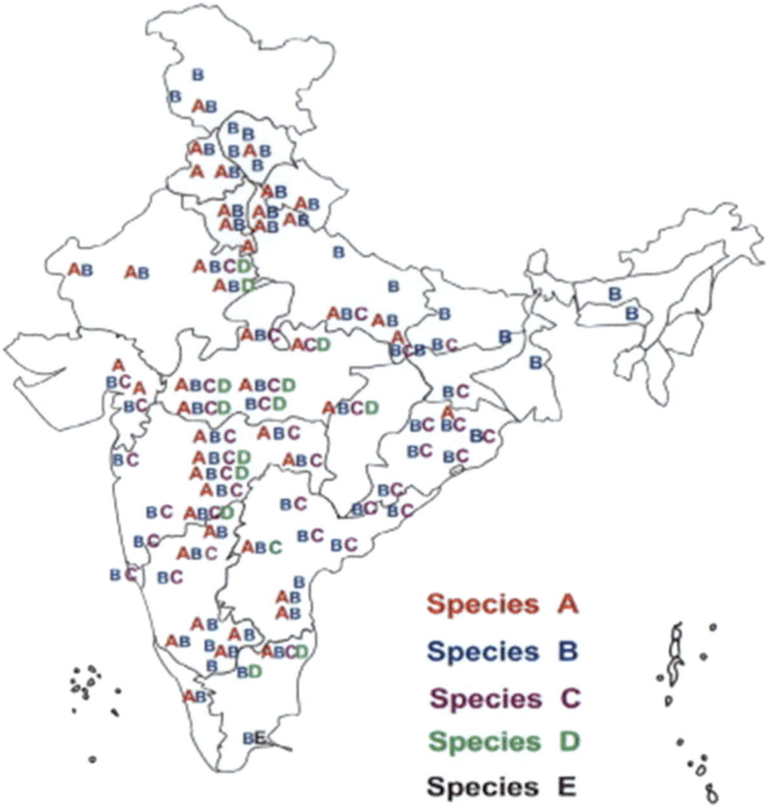

Fig. 15.2 Distribution of *An. culicifacies* sibling species in India (Source: NIMR, New Delhi)

sibling species to another. Among malaria vectors species, E is a very efficient vector, and its populations are sympatric with species B, as example in Rameswaram island [5] and Sri Lanka [6, 7].

Similarly *An. fluviatilis* is a complex of three sibling species designated as S, T and U [8]. Vector incrimination studies revealed that *An. culicifacies* B is a nonvector species, whereas other sibling species are malaria vectors, of which species E is a very efficient vector. The breeding habitats and response to insecticides also vary according to the sibling species. *An. culicifacies* transmits malaria in peri-urban areas along with *An. stephensi*. This is a vector of urban malaria with type form inhabiting the towns, and the intermediate and mysorensis inhabit the peri-urban and rural areas. Type form is a well-established malaria vector [9]. Initially the trials with DDT spraying to control malaria proved that DDT spraying would successfully eradicate malaria, but within the first 5 years of spraying, it became clear that eventually vector would develop resistance, and replacement of insecticides would result in the double and multiple resistance in the major malaria vector. Wherever resistance to DDT developed, dieldrin was sprayed, but vectors developed resistance to dieldrin also, and therefore HCH was sprayed. As the resistance developed, malaria continued to show resurgence and led to replacement of insecticide from HCH to malathion and then to synthetic pyrethroids. Malaria control is at the crossroads awaiting for the innovative interventions, but nothing seem to be emerging. Insecticide treated nets or LLINs gave hope of malaria control, but there are indications that vectors would develop widespread resistance to the chemicals used in the treatment of nets [10]. Obviously this strategy is good but only for a short while and not a permanent solution to the problem of malaria control. *An. culicifacies* is a major malaria vector and produces 65 to 70% malaria cases each year. The control so far has been transient. Control of *An. culicifacies*-transmitted malaria costs the country nearly 35% health budget of the Government of India and much higher allocation by the state governments. It has become obvious that this strategy cannot give a lasting solution. The basic need is the integrated vector management based on ecosystem approach. The breeding of *An. culicifacies* can be divided into several ecosystems, and a strategy may be developed for each ecosystem.

15.3 Link to Water

Mosquitoes breed in water to complete their aquatic life of eggs, larvae (I to IV instar) and pupae. They complete their aquatic life in about 7–15 days depending on the atmospheric temperature. Adult mosquitoes of both sexes emerge from the pupae and disperse in the environment to feed, mate and reproduce and thus complete their life cycle. Male mosquitoes typically live for a week or 2 weeks, and females may live for 4–6 weeks or more. Mosquitoes after emergence mate and disperse in surrounding areas in search of blood meal. Each mosquito species has its own preference for a blood meal, e.g. from warm-blooded animals to human beings; some beings are indiscriminate feeders. Females give rise to eggs which are laid on water surface singly (*Anopheles*) or rafts (*Culex*). Water quality influences the attraction of female mosquitoes to lay eggs. In general, clean water such as from the rains or rivers attracts the breeding of *Anopheles* and *Aedes* mosquitoes, whereas polluted water attracts *Culex* mosquitoes. Major mosquito breeding habitats for malaria vectors are as under *An. culicifacies*, the vector of rural malaria prefers stagnant water on the ground, seepage, rainwater, stored water for domestic use, agricultural field, wells, etc.; *An. stephensi*, malaria vector of urban and peri-urban malaria prefers stagnant rainwater, leaking water supply, water storage tanks, water holding receptacles, solid waste, domestic water, etc.; *An. fluviatilis*, malaria vector in plains and foothills prefers slow-flowing streams in plains and hills, rice fields, rainwater, etc.; *An. minimus*, malaria vector in the degraded forest breeds in slow-flowing streams predominantly in the north-eastern region and Orissa; *An. baimaii (=dirus)*, malaria vector in the north-eastern region particularly breeds in deep jungles in small rainwater collection, elephant footprints, and hoofprints; and *An. sundaicus*, malaria vector in Andman and Nicobar islands breeds in brackish water, wells with brackish water and a sibling species *An. sundaicus* breeds in sweetwater. Varieties of other mosquitoes which are encountered in the urban and rural areas that transmit deadly diseases breed in receptacles and discarded containers. These are *Aedes aegypti* which bred in containers' small water collections in the urban areas, desert coolers, coconut shells, discarded tins and pots, etc. *Aedes albopictus* principally breeds in forests in stagnant water, tree holes and leaf axils and often migrates to human habitations. It has been incriminated in the transmission of dengue and chikungunya fever along with or independently with *Aedes aegypti*. *Culex quinquefasciatus* mosquito, the vector for lymphatic filariasis, breeds in polluted water in open drains, disused wells, gutters and stagnant polluted water all over the rural and urban areas. This mosquito is the major cause of mosquito nuisance all over the country. Culicine mosquitoes mainly *Culex vishnui* group (*Culex tritaeniorhynchus*, *Culex vishnui* and *Culex pseudovishnui*) are the chief vectors of JE. A very large number of water bodies are found all over the rural and urban areas. Most of these water bodies produce huge mosquito population including disease vectors [11]. Some of the most common sources of mosquito production and disease transmission are (a) irrigation, (b) rainfall, (c) water supply and (d) marshy and low-lying lands, lakes, water bodies like the ponds, ditches, seepage, etc.

15.4 Malaria Ecotypes in India

At the time of launching of the NMCP in 1953, all malaria was rural malaria. With the progress in malaria eradication, malaria cases sharply declined in rural areas, and malaria was seen on its way out. At that time there was very little urban malaria, e.g. in Bombay (now Mumbai). In 1958 National Malaria Eradication Programme (NMEP) was launched, and urban malaria problem was not included for eradication. Small numbers of cases were to be tackled by the local governments. As the NMEP progressed towards malaria eradication, malaria cases were seen rising in the urban areas. The condition of malaria continued to deteriorate in urban areas so that in 1970–1971 urban malaria scheme was launched in 131 towns in the country [12]. Urban malaria continued to explode and occupied towns with piped water supply and water storage tanks, etc. the preferred place for the breeding of *An. stephensi*. Urban pull continued to draw population from the rural areas. Eventually almost all towns continued to expand. New settlements were unauthorized, lacked planning and poor to nil drainage; water supply and sanitation were scanty. On an average 25% population in urban areas now live in slums. Two vector systems operate, viz. *An. stephensi* and *An. culicifacies*. Developments under the 5-year plans to improve the national economy led to the creation of several more malaria ecotypes [13]. These are the forest malaria, in the degraded areas with *An. fluviatilis*, forests in the plains with *An. culicifacies*, degraded and foothills of the forests with streams *An. minimus* and deep jungles with *An. baimaii* as the main vector. Some coastal areas were colonized by the *An. sundaicus*, and in the wake of ecological changes, this vector has retracted from the mainland and still remains a major malaria vector in the A and N island [14]. All these ecotypes are broadly classified on the basis of terrain, vegetation cover, water quality, drainage and anthropogenic factors. It is noteworthy to mention that within these ecotypes, several sub-ecotypes operate.

Malaria controls in these sub-ecotypes require specialized training and experience. Integrated eco-based approach is required in vector control for successful attack on malaria on sustainable basis.

15.5 Integrated Eco-based Approach

15.5.1 Classical Examples

Pontine Marshland Near Rome Pontine marshland was inhabitable, and any attempt to move to that area was fatal. The vast land could not be colonized and remained barren. Benito Mussolini in 1932 organized drainage and cleared the mosquito breeding ground. The new settlements flourished and became a source of wealth generation. Mussolini thus became national hero. During the Second World War, Mussolini developed differences with Adolf Hitler. In 1943 Hitler ordered to fill the Pontine marshes with sea water and destroyed the pumps used for drainage. Malaria returned with vengeance [15]. Much after the World War II drainage in Pontine marshes was restored, and the land again became productive and with it the prosperity returned. Today Pontine marshland is very precious with industrial development, agriculture and modern settlements [16].

Panama Canal Construction of Panama Canal could not be undertaken due to high fatality rates of labour. Malaria and yellow fever were the most fatal diseases prevalent and required to be controlled before any work could be started. Mosquito breeding was everywhere, and there was no technology to control mosquito populations. The only way to overcome the problem was major earth work, drainage, integrated mosquito control and clearance of the marshes. This work was taken up on war footing and gradually with the disappearance of mosquito production sites malaria relented. It was then possible to construct the Panama Canal (1905–1910). The construction of Panama Canal has provided immense opportunities of trade between the nations, and the canal itself has become a major source of revenue to the host country [17].

Tennessee Valley The entire Tennessee valley was uninhabitable, and small population living in that land suffered from perennial malaria. The area had huge potential to transform the local economy but for the ravages of malaria. Through the intervention of the President of the USA establishing Tennessee Valley Authority (TVA), a special programme to reclaim the valley was evolved. Vector control was based on the elimination of mosquito breeding sites. In the reservoirs water level was changed to kill the eggs that stick to the walls of the reservoir [18].

15.5.2 Case Studies

In addition to the above classical examples of malaria control by environmental management methods, below are some examples from India where knowledge of terrain ecology and vector biology was applied in malaria control.

An. culicifacies Silk is a very precious commodity for the farming community in Karnataka. *An. culicifacies* is the major malaria vector which breeds in streams which have huge boulders. These streams go around villages providing a source of water for the villagers, their animals and wildlife. *An. culicifacies* mainly breeds in streams and wells and cement tanks. Malaria was rampant; in almost all houses, there were cases of fever diagnosed as malaria, both Pv and Pf, and deaths due to malaria were common in villages. Malaria control was not achievable as the farmers did not allow any spraying due to the possibility of harmful effects on the silk-worms. Mosquito fauna surveys revealed that *An. culicifacies* was the vector of malaria. Streams had big boulders, and larval control in the flowing streams was very difficult. Studies on the sibling complex revealed that species B breeds in the streams and species A in the wells and other habitats. The breeding in the streams was ignored, and in wells and other places, fishes were released to control mosquito breeding. The decline in populations of *An. culicifacies* species A resulted in the disappearance of malaria. In Karnataka Guppy and *Gambusia* fishes are very successful in malaria control [19].

An. stephensi Malaria in Bombay (=Mumbai) was a serious problem. There was intense breeding of *An. stephensi* in wells, water stored for domestic use in a variety of tanks, leaking water supplies, stagnant water ponds, rainwater collection on roof tops, etc. Covell (1928) designed an integrated malaria control strategy and demonstrated malaria control on sustainable basis [20]. The methods developed by Covell are still the same, and if applied in an integrated fashion produce lasting control. The strategy mainly included capping of all wells, mosquito proofing of water storage tanks, design for the mosquito proofing of overhead tanks, de-weeding and de-silting operations at regular intervals, legislative measures to prevent mosquito breeding in buildings, building bylaws and strict monitoring of all potential breeding sites all over Mumbai. These methods when applied in any urban area lead to mosquito control including the vector of urban malaria *An. stephensi*.

An. sundaicus (Island Malaria) In Nicobar island of A and Nicobar group of islands, malaria was the most common infection, and there were frequent cases of deaths. Vector surveys revealed that *An. sundaicus* is the only malaria vector. Mosquito breeding surveys revealed that *An. sundaicus* was breeding in brackish water in the creeks all along the coastline and the wells. DDT spraying for long did not eradicate malaria which was returning with vengeance year after year. Since mosquito breeding was in the brackish water produced by the mixing of rainwater and salt water in the creeks, one way sluice gates were installed to prevent the mixing of salt water. *An. sundaicus* breeding terminated automatically. In addition

Gambusia fishes were released in the wells. Thus sluice gates and *Gambusia* fishes combination completely controlled *An. sundaicus* breeding and malaria transmitted by *An. sundaicus* [21].

An. fluviatilis In the background of major ecological changes related to agricultural practices, it is essential that agriculture sector should be closely associated in the planning of malaria control. Land use pattern can mitigate many situations that otherwise would be receptive and vulnerable to malaria. This is an area of great opportunity and should not be missed. As, for example, Malnad region in southern India (Karnataka-Kerala) measuring about 50,000 sq. km was hot bed of malaria. Malnad forests were ideal terrain for coffee plantation. *An. fluviatilis* was the major malaria vector in Malnad. *An. fluviatilis* used to breed profusely in perennial streams in the forests in Malnad region. These streams originated from the rainwater were absorbed by the thick fallen vegetation cover. Malaria eradication became possible with the arrival of DDT. Malaria wasteland of Malnad was converted to coffee plantations. Jungle clearance gradually removed thick litter from the ground, and eventually seepage was eliminated. The terrain was very suitable for the breeding of *An. fluviatilis*, but with the disappearance of seepage, the breeding of *An. fluviatilis* has stopped. Since there was no other vector of malaria in Malnad, with the disappearance of seepage, Malnad became free of malaria and had become healthy region generating wealth from coffee plantation (Kalra, personal communication).

An. minimus and An. dirus (=baimaii) During the Second World War, British Indian forces had to fight the Japanese troops in Assam. Malaria transmission in the Assam war theatre was intense and perennial. Troops of both sides had to fight in malaria infested jungles. Japanese adopted the policy of replacement of sick troops and thus lost substantial fighting force. The British Indian army adopted the strategy of drainage in areas where the troop movement was planned. Indian side in Assam made the area malaria free. The technology adopted made the roads, railway station and military camps free of mosquitoes and malaria.

15.6 Integrated Vector Management

WHO describes integrated vector management (IVM) as 'a rational decision-making process for the optimal use of resources in the management of vector populations, so as to reduce or interrupt transmission of vector-borne diseases'. IVM includes five key elements: [1] evidence-based decision-making, [2] integrated approaches, [3] collaboration within the health sector and with other sectors, [4] advocacy, social mobilization and legislation and [5] capacity building [22], as shown in Fig. 15.1. The World Health Organization adopted the IVM for the control of vector-borne diseases in all regions of the world in 2004. The application of IVM in malaria control would lead to the optimum utilization of a mix of methods to improve the cost-effectiveness, environmental soundness and sustainability of interventions in the control of vector-borne diseases. The application of IVM in malaria control would require local knowledge of vectors, ecology, environmental considerations and selection of proven vector control methods to reduce or control vector-borne diseases at the acceptable levels. In doing so a range of vector control options would be considered for application singly or in combination synergistically. The task of control would require active involvement of the health and non-health sectors and engagement of the communities, social groups and stakeholders. Finally management of insecticide resistance, the use of other insecticides in a rational manner without any injurious consequences, is to be considered. IVM will give due consideration to the legislative framework to bring harmony in the integration of various techniques. Monitoring of interventions and consequences of control would be required for any mid-course improvement. IVM [23] would be implemented with the existing infrastructure and strengthened during implementation by training in vector control.

15.7 Concluding Remarks

Drainage India is a vast country, and poor drainage in rural and urban areas is an enormous problem causing a number of communicable diseases. The poor drainage, stagnant pools of water and growth of vegetation, has been the source of mosquito populations and prevalence of vector-borne diseases. Malaria has been the main disease afflicting the societies. Anywhere such poor drainage conditions were present; malaria was rampant as the wasteland remained an important source of mosquito production. Some of these areas, e.g. UP Terai, Malnad, were considered death traps. No development was possible unless the land was made habitable by clearing the mosquito breeding grounds of disease vectors. Land levelling, filling, channelling water bodies, etc. would be major works involved. While doing so, we should be careful to preserve the environment and avoid unnecessary interference in name of development. After the drainage and levelling of the land, other vector control methods can be deployed in an integrated fashion involving most appropriate and suitable strategy. Insecticides would play an important role, but their usage should be avoided wherever feasible. If a good job has been done prior to colonization, it is likely that vector invasion would be stopped without requiring chemical intervention. Detailed maps should be prepared and shared with the community who are likely to settle down in these lands. The maps should provide complete information on the various ecotypes requiring periodical interventions. We give below some practical examples of how the malaria wastelands were converted to productive land that transformed the country's economy. The control of *An. culicifacies*-transmitted malaria is the most formidable task. Wherever water is taken for agriculture or drinking purposes, it should be mandatory to provide necessary infrastructure to prevent water stagnation and seepage during and after the works are complete. In addition provision of adequate maintenance must be made, and all vulnerable sites should be inspected to prevent any water stagnation. De-silting operations must be completed before the rainy season for the free flow to water. A good drainage will ensure lasting protection from mosquito nuisance and prevention from diseases.

References

1. Sharma VP (2005) Malaria and other vector borne diseases in South Asia. In: Menon MGK, Sharma VP (eds) Sustainable management of water resources: emerging science and technology issues in South Asia. SCOPE XII General Assembly of SCOPE. February 7–11, 2005. Indian national Science Academy, Bahadur Shah Zafar Marg, New Delhi-110002. pp 337–358
2. Sharma VP (2012) Continuing challenge of malaria in India. Curr Sci 102(5):678–682
3. Malhotra MS, Shukla RP, Sharma VP (1985) Studies on the incidence of malaria in gadarpur town of U.P. Terai. Indian J Malariol 22(1):57–60
4. Subbarao SK, Adak T, Vasantha K, Joshi H, Raghuvendra K, Cochrane AH, Nussenzweig RS, Sharma VP (1988) Susceptibility of *Anopheles culicifacies* species a and B to *Plasmodium vivax* and *Plasmodium falciparum* as detected by immunoradiometric assay. Trans R Soc Trop Med Hyg 82:394–397
5. Kar I, Subbarao SK, Eapen A, Ravindran J, Satyanarayana TS, Raghvendra K, Nanda N, Sharma VP (1999) Evidence for a new vector species within the *Anopheles culicifacies* complex (diptera: culicidae). J Med Entomol 36:595–600
6. Jude PJ, Dharshini S, Vinobaba M, Sinnathamby S, N. Ramasamy R. (2010) Anopheles culicifacies breeding in brackish waters in Sri Lanka and implications for malaria control. Malar J 9:106
7. Surendra SN, Abhayawardana TA, De Silva BG, Ramasamy R, Ramasamy MS (2002) *Anopheles culicifacies* Y-chromosome dimorphism indicates sibling species (B and E) with different malaria vector potential in Sri Lanka. Med Vet Entomol 14:437–440
8. Subbarao SK, Nanda N, Vasantha K, Dua VK, Malhotra MS, Yadav RS, Sharma VP (1993) Cytogenetic evidence for three sibling species in *Anopheles fluviatilis* (diptera: culicidae). Ann Entomol Soc Am 87(1):116–121
9. Rao TR (1984) The anophelines of India, Revised edn. Published by the Malaria Research Centre, New Delhi
10. Becker ND, Marija Z, Clive B, Minoo M, Christine D, Achim K (2010) Mosquitoes and their control. Springer, 577 pages
11. Sharma VP (2012b) Water mosquitoes and malaria. In: Water and health. Springer, New Delhi. In Press
12. NVBDCP Website (2012)
13. Pattanayak S, Sharma VP, Kalra NL, Orlov VS, Sharma RS (1994) Malaria paradigms in India and control strategies. Indian J Malariol 31:141–195
14. Sharma VP, Prasttisuk P, Kondrashin AV (1991) Magnitude of the forest related malaria in the WHO South East Asia Region. In: Sharma VP, Kondrashin AV (eds) Forest malaria in South East Asia. Proceedings of an informal consultative meeting. Malaria Research Centre (ICMR) Delhi-110054, India. pp 29–53
15. Frost RS (1934) The reclamation of Pontine marshes. Geogr Rev 24(4):584–595
16. Caprotti F (2006) Malaria and technological networks: medical geography in the Pontine marshes, Italy, in the 1930s. Geogr J 172(2):145–155
17. Schlesinger AT (1999) Building the panama canal. Chronicles from national geographic. Chelsea House Publishers, Philadelphia
18. Hanson K, Catherine G, Jo L, Sylvia M, David B, Anne M (1984) Vector control using insecticides. Intech, Geneva
19. Ghosh SK, Tiwari SN, Sathyanarayan TS, Sampath TRR, Sharma VP, Nanda N et al (2004) Larvivorous fish in wells target the malaria vector sibling species of the Anopheles culicifacies complex in villages in Karnataka, India. Trans R SocTrop Med Hyg 99:101–105
20. Covell G (1928) Malaria in Bombay. Central Government Press, Bombay
21. Das MK, Nagpal BN, Ansari MA (2002) Mosquito fauna and breeding habitats of anophelines in little Andaman Island, andaman and Nicobar Islands, India. Indian J Malariol 39:83–95
22. Beier JC, Keating J, Githure JI, Macdonald MB, Impoinvil DE, Novak RJ (2008) Integrated vector management for malaria control. Malar J 11(7 Suppl 1):S4

23. van den Berg H, Mutero CM, Ichimori K (2012) Guidance on policy-making for integrated vector management. WHO/HTM/NTD/VEM/2012.2. World Health Organization, Geneva

Dr. Vinod Prakash Sharma, MSc (1960), DPhil (1964), DSc (1979), all from Allahabad University, is postdoctoral research associate of the University of Notre Dame and Purdue University, USA (1964–1968); pool officer of the Forest Research Institute, Dehradun (1969–1970); senior scientist of WHO/ICMR Research Unit on the Genetic Control of Mosquitoes, New Delhi (1970–1975); deputy director of Vector Control Research Centre/Malaria Research Centre (1975–1982); founder director of the National Institute of Malaria Research (1962–1998); and additional director general of the Indian Council of Medical Research (1998). He is FNA, FASc, FNASc, FAMS, FTWAS and fellow of several other learned societies. He has specialized in malaria and vector biology. He has 40 years of research and field experience, published 400 scientific papers and book on *Anopheles* and edited 14 books. Under his leadership, NIMR had earned international reputation of excellence in malaria research and control and established linkages with leading national and international laboratories. He conceptualized the bioenvironmental malaria control strategy as an alternative to spraying and successfully demonstrated malaria control in various eco-epidemiological settings. Dr. Sharma's work has become legendry. As a result of his indefatigable efforts, today we have technologies to fight malaria which are safe and cost-effective, produce sustainable impact and are free from environmental hazards produced by DDT and other insecticides. His work has been recognized nationally and internationally. He was invited to work on various WHO panels since 1980s. He was chair of the WHO/UNEP/FAO Panel of Experts on Environmental Management (PEEM) for Vector Control (two 5-year terms). WHO invited him to write/review GFATM proposals for SEA Regional countries, i.e. Indonesia, Timor Leste, Myanmar, Bhutan, Bangladesh and India, and for Sierra Leone by the NGO BRAC. He was the principal architect for writing World Bank and other projects for the Indian programme on malaria control. He chaired review committees of the NMEP, Nepal, Bhutan and Thailand, and is actively involved in the evaluation of countries capabilities to proceed towards malaria elimination. He was invited in WHO (Geneva and New Delhi) meetings as temporary adviser. He represented India on the Scientific Committee on the Problems of the Environment (SCOPE), Paris, for nearly 20 years and later was elected as executive member of the SCOPE. In that capacity he coordinated the "Health and Environment" cluster of the SCOPE. He was president of the National Academy of Sciences, India (1999–2000); Indian Society for Parasitology (1993–1997); and National Academy of Vector and Vector Borne Diseases (1979–2008); chief editor of the *Journal of Parasitic Diseases* and *Indian Journal of Malariology*; council member of the International Congress of Entomology, UK (two terms); and member of the WHO Expert Committee on Malaria, Geneva, still continuing. He is a recipient of several national and international awards/honours, e.g. WHO Darling Foundation Prize (1999); ICMR Ambedkar and MOT Iyengar awards; Lifetime Achievement Award; Green Scientist Award; G.P. Chatterjee Memorial Award; Ranbaxy award; FICCI award; Om Prakash Bhasin Award; Vasvik awards; distinguished parasitologist recognition by the World Federation of Parasitologists (2010); Gujar Mal Modi Award (2013); Vigyan Vibhuti award by the Uttrakhand State Council for Science and Technology (2013); Meghnad Saha Distinguished Fellow (2005–2010); NASI Distinguished Professor and ICMR Chair in Public Health Research (2010–2015) at the Indian Institute of Technology, New Delhi; and Padma Shri (1992) and Padma Bhushan (2014) awarded by the President of India.

Health Impact of Bacteriological and Chemical Contamination of Drinking Water with Special Reference to Arsenic

16

D.N. Guha Mazumder

An adequate supply of safe drinking water is essential for a healthy life, but waterborne disease is still a major cause of death in various parts of the world. The drinking water quality and possible associated health risks vary in different parts of the world with some regions showing, for example, contamination of drinking water by pathogenic agents or high levels of arsenic and fluoride, whereas these are very low and no problem elsewhere. Water may be polluted with pathogenic bacteria by excreta or sewage, which is certain to contain pathogenic microorganisms. Further, microbial contamination of water can occur in unhygienic storage, poor hygiene practice of individuals and by flies carrying the pathogenic organisms. Chemical contamination in water can be either naturally occurring or introduced by human activity and can have serious health effects.

16.1 Microbial Contamination

Gastrointestinal infections commonly produce diarrhea caused by a variety of pathogens, including bacteria, viruses, and protozoa. But, only a handful of organisms are responsible for most acute cases of childhood diarrhea [1]. *Rotavirus*, a leading cause of acute diarrhea, is responsible for about 40% of admissions in hospital due to diarrhea among children under 5 worldwide. *E. coli*, *Shigella*, *Campylobacter*, and *Salmonella*, along with *V. cholera*, are the major bacterial pathogens which are involved during epidemics [1]. Other agents involved are *Campylobacter jejuni*, enterotoxigenic *Escherichia coli*, and possibly enteropathogenic *E. coli*, *Aeromonas* spp., *V. cholerae* O139, enterotoxigenic *Bacteroides fragilis*, *Clostridium difficile*, and *Cryptosporidium parvum*. Chlorination of water can control all except the latter, but recontamination of treated water is a great problem.

D.N. Guha Mazumder (✉)
DNGM Research Foundation, Kolkata 700053, India
e-mail: dngmazumder@gmail.com

K.J. Nath, V.P. Sharma (eds.), *Water and Sanitation in the New Millennium*,
DOI 10.1007/978-81-322-3745-7_16

Often cholera is thought of as a major cause of child deaths due to diarrhea; however most cases occur among adults and older children. Over and above the traditional pathogens (helminths, Entamoeba histolytica, Giardia lamblia, hepatitis A and E viruses), other emerging issues in adults are various *Enteroviruses*, *C. jejuni*, and *H. pylori* [2]. In the tropical countries like India, hepatitis, cholera, dysentery, and typhoid are the more common waterborne diseases that affect large populations.

A study which was carried out using a standardized protocol in five hospitals in China, India, Mexico, Myanmar, and Pakistan information on major causes of acute diarrhea in children aged 0–35 months was available from a total of 3640 cases of diarrhea and 3279 age- and sex-matched controls. About 60% of the patients were aged less than 1 year and 60% were male. An enteric pathogen was isolated in 66% of the cases and in 30% of the controls. The pathogens most strongly associated with disease were *Rotavirus* (16% of cases, 2% of controls), *Shigella* spp. (11% of cases, 1% of controls), and enterotoxigenic *Escherichia coli* (16% of cases, 5% of controls) in all the study centers. Commonest pathogen isolated among 6–11 months old was *Rotavirus*, accounting for 20% of all cases in this age group; 71% of all *Rotavirus* episodes occurred during the first year of life. Findings from these studies suggest that microbe-specific intervention strategies for the control of childhood diarrheal diseases in developing countries should focus on *Rotavirus*, *Shigella* spp., and enterotoxigenic *E. coil* [3].

Diarrhea is the second leading cause of death among children under five globally. About one in five child deaths – about 1.5 million each year – is due to diarrhea. Diarrhea kills more young children than AIDS, malaria, and measles combined. Total number of child death due to diarrhea in India is 386,600 annually [4]. New approaches to increase demand to stop open defecation have proven more effective in regard to community-wide sanitation than previous strategies. It has been estimated that 88% of diarrheal deaths worldwide are caused by unsafe water, inadequate sanitation, and poor hygiene [5]. A number of interventions that work to reduce the number of diarrhea cases are water, sanitation, and hygiene programs. Disposing of human excreta in a sanitary manner, washing hands with soap, increasing access to safe water, improving water quality at the source, and treating household water and storing it safely are considered to be the most effective intervention for reducing occurrence of diarrhea [4].

Treatment Package

Two main elements, as laid out below, are focused as treatment package [6]:

1. Fluid replacement to prevent dehydration
2. Zinc treatment

Prevention Package

In the medium- to long-term goal, prevention package focuses on five main elements to reduce diarrhea; these are:

3. *Rotavirus* and measles vaccinations
4. Promotion of early and exclusive breastfeeding and vitamin A supplementation
5. Promotion of hand washing with soap
6. Improved water supply quantity and quality, including treatment and safe storage of household water
7. Community-wide sanitation promotion

16.2 Chemicals in Drinking Water

The chemical contaminations in drinking water are seldom high enough to cause acute health effects. They most commonly cause chronic health effects – effects that occur following long-term exposure to small amounts of a chemical. Examples of various chronic health effects of prolonged chemical toxicity are cancer, birth defects, organ damage, disorders of the nervous system, and damage to the immune system. Usually limited evidences are available relating to chronic health effects to specific drinking water contaminants. Frequently scientists predict the likely adverse effects of chemicals in drinking water using laboratory animal studies in the absence of exact scientific information. However, human data from clinical reports and epidemiological studies provide a lot of information, when available.

16.2.1 Fluoride

Varying amounts of fluoride are found naturally in water supplies. Benefits of dental health following the use of optimally fluoridated (<1 mg/L) water have been well documented; it causes reduction of the incidence of tooth decay. But mottling of teeth, called fluorosis, may occur following intake at concentrations over 1.5 milligrams per liter. Reports of occurrence of fluoride-contaminated groundwater have been available from many states of India, namely, Andhra Pradesh, Bihar, Tamil Nadu, Kerala, Maharashtra, Punjab, Haryana, Rajasthan, Gujrat, Uttar Pradesh, West Bengal, Assam, Orissa, Jharkhand, and Karnataka [7].

Fluorosis occurs in three forms: dental, skeletal, and nonskeletal fluorosis. Dental fluorosis will be visible from discoloration of permanent teeth aligned horizontally and/or discoloration in spots away from gums, mostly in children. In skeletal fluorosis, generalized bone and joint pain occur in mild cases. This is followed by stiffness of joints with restricted movement of spine and joints. Finally, flexion deformity develops in spine and joins. Nonskeletal fluorosis is characterized by muscle weakness, gastrointestinal symptoms like loss of appetites, pain abdomen, and anemia.

For confirmation of diagnosis, fluoride level in drinking water source and in urine and blood of suspected cases may be tested. Normal fluoride level in water, urine, and blood is up to 1 mg/L, 0.1 mg/L, and 0.02 mg/L, respectively [8].

16.2.2 Nitrate

Contamination of water supplies with nitrate can occur from many sources – chemical fertilizers, animal wastes, manure, and sewages. According to the information available, Tamil Nadu, Haryana, and Rajasthan are the few states in India having high nitrate content in water sources. Intake of nitrate-contaminated water can cause serious problem because in the digestive systems of human infants and in some livestock, nitrate taken through drinking water is converted to nitrite, a very toxic substance. Because of presence of certain bacteria in the digestive systems at birth, infants are extremely susceptible to acute nitrate poisoning. These bacteria change nitrate into toxic nitrite. Hemoglobin present in the blood carries oxygen. The nitrite reacts with hemoglobin in the bloodstream to form methemoglobin which cannot carry oxygen. As the infant suffers from methemoglobinemia, the circulating oxygen level in blood decreases. The condition is called blue baby disease, which is characterized by a bluish skin color, particularly around the eyes and mouth. This is a sign of respiratory failure, and the infant should be taken to the doctor immediately for treatment.

The digestive system is fully developed by the time the baby is 6 months old, and none of the nitrate-converting bacteria appear to remain in the intestine. In older children and adults, nitrate is generally absorbed and excreted. However prolonged intake of high level of nitrate can cause gastric problems in adults due to formation of carcinogenic nitrosamines [9].

16.2.3 Arsenic

16.2.3.1 Chronic Arsenic Toxicity

A major environmental health hazard throughout the world including India and Bangladesh has been happening due to chronic arsenic toxicity (arsenicosis) due to drinking of arsenic-contaminated groundwater. Significant arsenic contamination in groundwater was detected in the year 1983 in West Bengal, India, when some villagers were diagnosed to be suffering from arsenicosis due to drinking of arsenic-contaminated water. Currently, the arsenic contamination in groundwater has been detected in 79 blocks in 8 districts of West Bengal. The major districts affected are Malda, Murshidabad, Nadia, Burdwan, and North and South 24 Parganas (Fig. 16.1). It is conjectured that six million people are exposed to arsenic-contaminated groundwater (>50 μg/L). Occurrences of arsenic in groundwater have also been reported from other states in India, e.g., Bihar, Jharkhand, Chhattisgarh, Uttar Pradesh, and Assam [10].

Significant research on health effects of chronic arsenic toxicity in humans has been done in West Bengal during the last 28 years. The symptoms of chronic arsenic toxicity (arsenicosis) are dependent on the magnitude of the dose and duration of its exposure and occur insidiously. Further, the incidence of chronic arsenicosis in an affected population varies widely. Even, all members of an affected family do not show clinical symptoms of arsenicosis. Specific skin lesions characteristic of

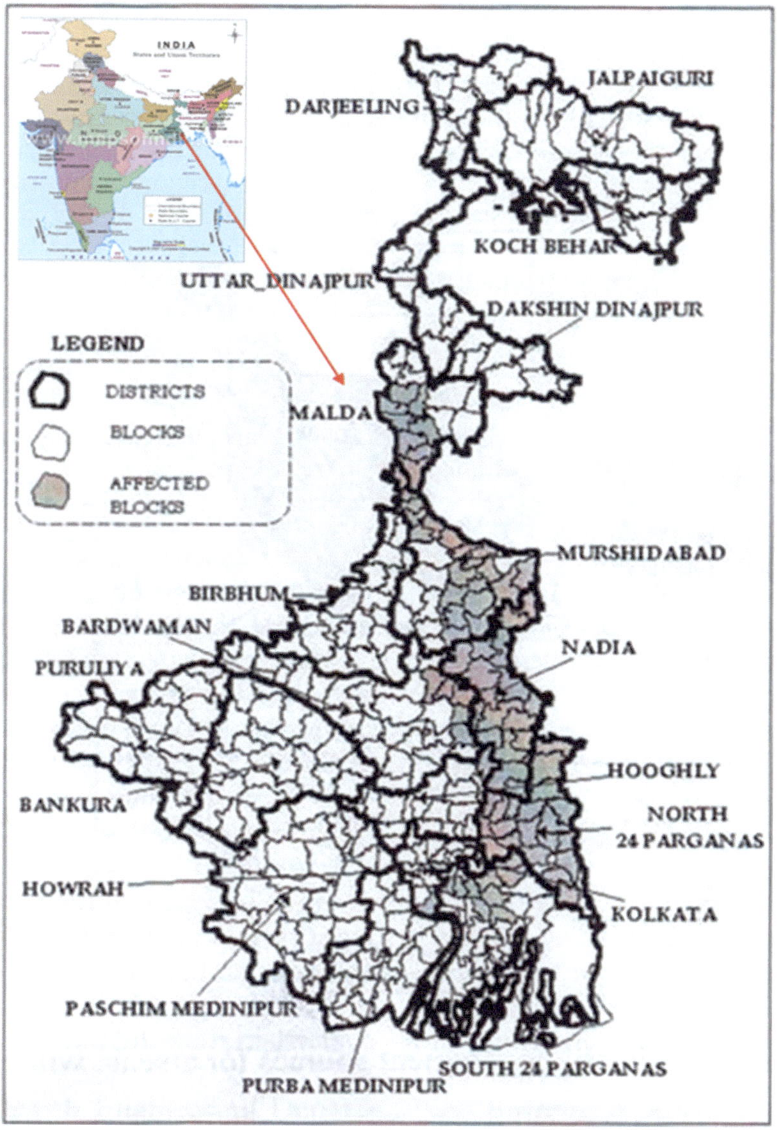

Fig. 16.1 Map of West Bengal showing arsenic-affected districts

chronic arsenic toxicity are pigmentation and keratosis. However, it was found to be associated with various systemic manifestations including cancer in West Bengal.

16.2.3.2 Skin Manifestations

Chronic arsenic toxicity causes pigmentation and keratosis, the characteristic specific skin lesions. The pigmentation of chronic arsenic toxicity is bilaterally distributed and commonly appears in a finely freckled "raindrop" pattern of pigmentation or depigmentation that is particularly pronounced on the trunk and extremities. Pigmentation occurs in different parts in the body, specially marked in non-exposed parts of the body, such as the trunk, buttocks, and thighs. Sometimes pigmentation may be blotchy and involve mucous membranes such as the under surface of the tongue or buccal mucosa. Arsenical hyperkeratosis occurs predominantly on the palms and the plantar aspect of the feet; however involvements of the dorsum of the extremities and the trunk have also been described. Most frequently they are found on the thenar eminence and the lateral borders of palms, fingers, and on the soles, heels, and toes. The involved skin might have an indurated, grit-like character that can be best appreciated by palpation in the early stage of disease; however, the lesions usually advance to form raised, punctate, 2–4 mm wartlike keratosis that are readily visible [11]. To assess the dose response relationship and arsenical skin lesion, the first population-based survey was carried out in the district of South 24 Parganas, West Bengal, on 7683 participants (4093 female and 3590 male) out of a total population of 150,457 with individual arsenic exposure data from drinking water sources. Water samples, collected from all recruited households of the participants, were tested for arsenic by flow-injection hydride generation atomic absorption spectrophotometry. The tube well arsenic concentration varied from non-detectable to 3200 μg/L. The overall prevalence rates for keratosis and hyperpigmentation were 3.64 and 8.82 per 100, respectively. Prevalence of keratosis and hyperpigmentation was determined by water arsenic level. A clear relationship was observed between water levels of arsenic and keratosis and hyperpigmentation. Age-adjusted prevalence of keratosis and hyperpigmentation was strongly related to water arsenic levels, increasing from 0 and 0.3 in the lowest exposure level (<50 μg/L) to 8.3 and 11.5 per 100, respectively, for females drinking water containing >800 μg/L and raising from 0.2 and 0.4 per 100 in the lowest exposure category to 10.7 and 22.7 per 100, respectively, for males in the highest exposure level (>800 μg /L). The prevalence of skin lesion was also assessed by daily dose per body weight (μg/Kg/day).This showed that men had roughly two to three times higher prevalence of keratosis and pigmentation compared to those for women apparently taking the same dose of arsenic from drinking water [12] (Fig. 16.2).

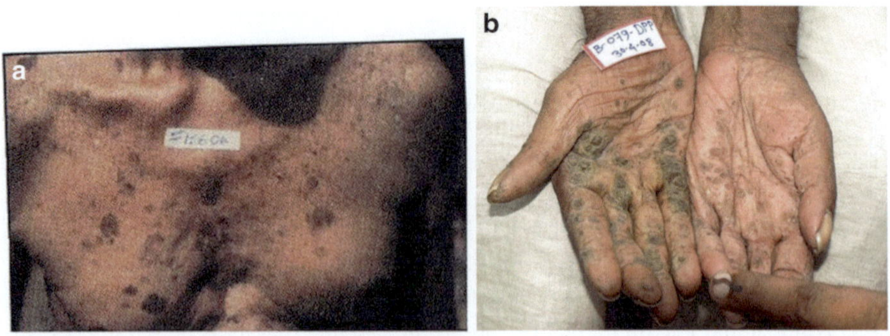

Fig. 16.2 (**a**) Severe arsenical pigmentation. (**b**) Severe arsenical keratosis

16.2.3.3 Systemic Manifestations

Over and above skin lesions, chronic arsenic toxicity is associated with various systemic manifestations (Table 16.1). A scientific epidemiological study was carried out by adopting stratified multistage design in Nadia district, one of the arsenic-affected districts of West Bengal to ascertain the global disease impact of arsenicosis. People in all its 17 blocks of the district are affected by groundwater arsenic contamination. Chronic lung disease and peripheral neuropathy were found to be the major systemic manifestations in our study. However other systemic manifestations, less in numbers, were also present. [13]. In Taiwan where blackfoot disease (a form of peripheral vascular disease) is endemic due to drinking of arsenic-contaminated water from artisan well water; other systemic diseases like hypertension, ischemic heart disease, and cerebrovascular disease have been mostly reported [14]. Though systemic manifestations occur mostly in association with arsenical skin lesions, symptoms also occur in significantly higher number of cases in arsenic-exposed people even in the absence of arsenical skin disease compared to arsenic-unexposed people [15].

Table 16.1 Data showing clinical features of chronic arsenic exposure in epidemiological study in Nadia, West Bengal [23]

	Population-based study in Nadia in 2006–2007 in arsenic-endemic region (n = 10,469)				
Parameter	(Cases) with arsenical skin disease		(Control) without skin disease		p-value
(Mean ± SD)	(n = 1616)		(n = 8853)		
Water arsenic exposure (μg/L)	87 ± 126		64 ± 101		<0.001
Age	53.36 ± 15.60		33.74. ± 15.99		<0.001
Sex	n	%	n	%	
Male	934	57.80	3213	36.29	<0.001
Female	682	42.20	5640	63.71	<0.001
Disease Symptoms					
1. Skin score					
Mild	1415	87.5			
Mod	187	11.5			
Severe	14	0.87			
2. Ch. bronchitis	127	7.86	52	0.59	<0.001
3. Breathlessness (COPD)	146	9.03	39	0.44	<0.001
4. Dyspepsia	67	4.15	77	0.87	<0.001
5. Ch. Diarrhea	19	1.18	15	0.17	<0.001
6. Liver (palpable)	5	0.31	0	0.00	<0.001
7. Ascites	3	0.19	1	0.01	<0.05
8. Pallor (anemia)	1	0.06	7	0.08	>0.05
9. Limb swelling (non-pitting edema)	4	0.25	2	0.02	<0.01
10. Per. neuritis	257	15.90	136	1.54	<0.001

16.2.3.4 Pregnancy Outcome

As few studies included individual assessment of arsenic concentrations in all water sources used during each pregnancy, limited information on pregnancy outcome and infant mortality in relation to arsenic levels in drinking water is available in literature. Pregnancy outcome and infant mortality study was conducted in West Bengal, India, retrospectively between the years 2001 and 2003, among 202 married women selected from a source population of 7683 having all the arsenic exposure data available throughout pregnancy. This study showed that exposure to high concentrations of arsenic during pregnancy increases the risk of stillbirth. However, there was no evidence of increased rates of spontaneous abortion and overall infant mortality [16].

16.2.3.5 Arsenicosis and Cancer

Arsenic-induced carcinogenicity in humans is based on the evidences obtained from epidemiological studies of cancer in relation to arsenic exposure through drinking water. On the basis of evaluation of data obtained from ecological studies, cohort studies, and case-control studies from many countries, the working group of International Agency for Research on Cancer observed that arsenic was potentially carcinogenic for skin cancer, urinary bladder cancer, and lung cancer due to chronic exposure to arsenic [10]. A total of 212 (4.35%) cases of skin cancer and 38 (0.78%) internal cancers were detected among 4865 cases of arsenicosis studied in arsenic-affected villages in West Bengal [17] (Table 16.2).

Laboratory criteria for determining exposure history of arsenicosis cases [18] (WHO 2005).

Testing water: Intake of drinking water with an arsenic concentration in excess of prevailing national standards for at least 6 months, Country Standard (Asia-Pacific region), 0.0 5 mg/L; WHO Standard, 0.01 mg/L.

Testing biomarker: If results of arsenic concentration of previously consumed drinking water are unavailable, an elevated level of arsenic in hair (> 1 mg/Kg of hair) or in nail clippings (> 1.5 mg/Kg of nail) may serve as presumptive evidence of elevated arsenic exposure.

Table 16.2 Observations on occurrence of cancer associated with arsenic exposure in West Bengal

Symptoms	Guha Mazumder D.N, 1999, Hospital-based study [n = 248] [unpublished data] %	Chatterjee D, 2002, Hospital-based study [n = 470/122] [OPD/admitted cases] [unpublished data] %	Saha KC, 2003 Village-based study [n = 4865] %
Skin cancer	02.02	4.04/13.11	04.35
Internal cancer	0.4	0.42/3.9	0.78

16.3 Conclusion

Results of various studies done on health effect of chronic arsenic toxicity during last 28 years in West Bengal showed that chronic lung disease, neuropathy, and chronic liver disease over and above skin manifestations are the major cause of morbidity, while chronic lung disease and cancer are major causes of mortality among the arsenic-exposed population. The main emphasis in management of arsenicosis needs to be given on intake of arsenic-free water and high-protein diet. There is lack of awareness among the people regarding the danger of arsenic and lot of misconception on skin manifestations cause social problem. Important risk factor for disease manifestations is poverty associated with lack of protein and fiber intake in the diet. Health effect due to arsenic exposure through diet is currently being investigated as there are reports of high arsenic content in the rice in the region.

References

1. World Health Organization (1999) The Evolution of Diarrhoeal and Acute Respiratory Disease, Geneva. http://whqlibdoc.who.int/hq/1999/WHO_CHS_CAH_99.12.pdf. Accessed July 2012
2. Ashbolt NJ (2004) Microbial contamination of drinking water and disease outcomes in developing regions. Toxicology 198:229–238
3. Sima H, Guang ZL, Mathan MM, Mathew MM, Olarte J, Espejo R, Khin M, Ghafoor UMA, Khan MA, Sami Z, Sutton RG (1991) Etiology of acute diarrhoea among children in developing countries: a multicentre study in five Countries. Bull World Health Organ 69:549–555
4. The United Nations Children's Fund (UNICEF)/World Health Organization (WHO) (2009) Diarrhoea: why children are still dying and what can be done. UNICEF, New York
5. Black RE, Morris S, Bryce J (2003) Where and why are 10 million children dying every year? Lancet 361:2226–2234
6. United Nations Children's Fund and World Health Organization (2004) WHO/UNICEF joint statement: clinical management of acute diarrhoea. UNICEF, New York. http://www.afro.who.int/cah/documents/intervention/acute_diarrhoea_joint_statement.pdf. Accessed June 2009
7. Nanoti M. Fluoride: sources and treatment. Water quality management: South Asian Perspective vision 2005. International Life Sciences Institute-India, New Delhi, pp 105–117
8. Susheela AK. Adverse health effects of Fluoride in drinking water with special reference to children and remedial measures. *Ibid.* pp 118–134
9. Jain SK. Sources, treatment and health aspects of nitrates and total dissolved solids. *Ibid.* pp 135–154
10. IARC (2004) Some drinking-water Disinfectants and contaminants, including Arsenic. Monographs on the evaluation of carcinogenic risks to humans. Lyon, France, WHO 2004, 84: 61–96
11. Guha Mazumder DN (2001) Clinical aspects of chronic arsenic toxicity. J Assoc Physicians India 49:650–655
12. Guha Mazumder DN, Haque R, Ghosh N, De BK, Santra A, Chakraborty D, Smith AH (1998a) Arsenic levels in drinking water and the prevalence of skin lesions in West Bengal, India. Int J Epidemiol 27:871–877
13. Guha Mazumder DN, Ghosh A, Majumdar KK, Ghosh N, Saha C, Guha Mazumder RN (2010) Arsenic contamination of ground water and its health impact on population of district of Nadia, West Bengal, India. Indian J Community Med 35:331–338
14. NRC (National Research Council) (1999) Health effects of arsenic. In: Arsenic in drinking water. National Academic Press, Washington, DC, pp 83–149

15. Majumder KK, Guha Mazumder DN, Ghose N, Lahiri S (2009) Systemic manifestations in chronic arsenic toxicity in absence of skin lesions in West Bengal. Indian J Med Res 129:75–82
16. Von Ehrenstein OS, Guha Mazumder DN, Smith MH, Ghosh N, Yuan Y, Windham G, Ghosh A, Haque R, Lahiri S, Kalman D, Das S, Smith AH (2006) Pregnancy outcomes, infant mortality and arsenic in drinking water in West Bengal, India. Am J Epidemiol 163:662–669
17. Saha KC (2003) Saha's grading of arsenicosis progression and treatment. In: Chappell WR, Abernathy CO, Calderon RL, Thomas DJ (eds) Arsenic exposure and health effects V. Elsevier Science, Oxford, pp 391–414
18. WHO Technical Publication (2005) A field guide for detection, management and surveillance of arsenicosis cases. In: Caussy D (ed), WHO Regional Office for South East Asia, New Delhi, 30, pp 19–22

Dr. D.N. Guha Mazumder (born 1934), MD, FAMS, recipient of the Coats Medal (1996) from Calcutta University for most notable contribution in the preceding 7 years and Barclay Memorial Medal (2010) from the Asiatic Society for significant contribution in the field of medical science for India, is the director of DNGM Research Foundation, Kolkata, a charitable trust established by him (1999). He retired as professor and head of the Department of Medicine and Gastroenterology, Institute of Post Graduate Medical Education and Research, Kolkata (1996). He conducted research on various diseases like mycotoxicosis, diphtheria, smallpox, tetanus, acute diarrhoeal disease, tropical sprue, giardiasis, ascariasis and dyspepsia including *H. pylori* infection, non-cirrhotic portal fibrosis and viral hepatitis. Further, he has been carrying on studies on the problem of arsenic-related health hazard for more than 30 years. He received funding from the ICMR; CSIR; NIH; University of California, Berkeley, USA; UNICEF; and World Bank for undertaking research. He contributed on the diagnosis and treatment of chronic arsenic poisoning in the *United Nations Synthesis Report on Arsenic in Drinking Water* in 2000 and in the *IARC Monographs on the Evaluation of Carcinogenic Risks to Humans* in 2004. He published more than 140 papers in journals, monographs and books.

Sustainable Management of Water Resource in India: Feasibility of Linking the Rivers 17

Kalyan Rudra

17.1 Introduction

The history of the human civilisation is the story of struggle between man and nature, represented as interaction–response–interaction, where man never achieved absolute sovereignty but continued to alter the nature. The water played a critical role in shaping the civilisation. It quenched the thirst, helped to flourish agriculture, designed the layout of the city or village, set the limit of expansion, delineated the boundary of the country and rendered the gateway for trade and commerce. The contaminated water also spread diseases leading to deaths of millions of people, or the onrushing floods ravaged civilisations. The non-availability of water left extensive parts of the surface deserted. The aridity, human thirst, water control and political power are thus intimately related.

In the beginning of the new millennium, water has become a major issue of the Indian politics or means of consolidating power [10]. The water scarcity, not the abundance, deepens our anxiety. One-sixth of the country is declared as drought prone, but almost one-third of the country suffers from acute water shortage in every summer, and not less than 270 million people struggle for their livelihood in a parched condition. It is officially admitted that 40 million hectares of land or about 12% area of the country with a population of 60 million is prone to annual flooding. The structural measures like dams and embankments have failed to ensure protection against flood but have rather aggravated the situation in many areas [1, 10]. Water has started to play a pivotal role in the electoral politics of our country. The politicians, irrespective of their colour and belief, leave no stone unturned to exploit the situation. This reminds us that 'control over water has again and again provided an effective means of consolidating power within the human groups-led, that is, to the assertion by some people of power over other' [5].

Our country is known to be the gift of its unique river systems. The rivers have been flowing along the lines, which are unquestionably the best possible routes of

K. Rudra (✉)
West Bengal Pollution Control Board, Kolkata, West Bengal, India
e-mail: rudra.kalyan@gmail.com

K.J. Nath, V.P. Sharma (eds.), *Water and Sanitation in the New Millennium*,
DOI 10.1007/978-81-322-3745-7_17

flowing water. But the proposed plan is to large scale interbasin transfer water to satisfy the human need as if water has nothing to do in the ecosystem. The idea is to modify the delicately balanced natural system and introduce a structurally controlled hydraulic system to combat the twin problems of flood and drought (Fig. 17.1).

India is bestowed with 4000 km^3 of freshwater annually, from precipitation. Excluding the evaporation, infiltration and interception losses, the available surface water amounts to 1869 km^3 – 37% of which (i.e. 690 km^3) can be utilised for beneficial purposes in addition to 432 km^3 of replenishable groundwater. Thus, the total water available to meet our demand is estimated to be 1122 km^3 [12]. The annual per capita available water in India has steadily declined from 5177cu.m. in 1951 to about 1820 m^3. in 2001 and continues to dwindle further with the uninterrupted growth of population (see Tables 17.1 and 17.2). The total demand in this country will be equivalent to utilisable freshwater resource by the middle of the twenty-first century [3]. The official projection of demand of water in various sectors shows that

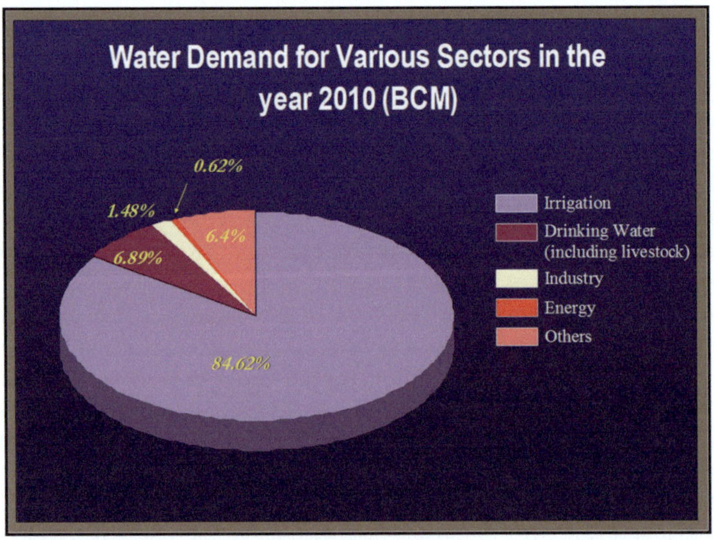

Fig. 17.1 Demand of water in different sectors

Table 17.1 Water availability in India (in billion cubic metres)

Annual Precipitation (including snow)		4000
Seasonal Rainfall (June to September)		3000
Average Annual Flow in Rivers		1869
Annual Utilisable Water Resources		
(a) Surface	690	
(b) Ground	432	
Total	1122	

Source: Ministry of water resource (Govt. of India)

Table 17.2 Per Capita water availability (in cubic metres)

Year	Population in million	Water availability
1951	361	5177
1955	395	4732
1991	846	2209
2001	1027	1820
2025	1394	1341
2050	1640	1140

Source: Ministry of water resource (Govt. of India)

the agriculture will continue to be the largest consumer of water (85%) even in 2010, whilst the demand for drinking and industry would be 6.65% and 6.45%, respectively. The paradox of the Indian water management is that Cherrapunji (Meghalaya), which gauges the highest rainfall in the world (more than 11,000 mm annually), now faces the problem of water shortage for 6 months in a year, whilst Alwar (Rajasthan) in spite of being located in the arid west has achieved remarkable success in rainwater harvesting. Despite ever-increasing investment and expenditure in water management, both the flood- and drought-prone areas have been expanding. The Government admits that 190 million people in this country still do not have access to safe drinking water. Again, 80% of the children suffer from waterborne diseases, and 0.70 million of them die each year. Moreover, 44 million people suffer from diseases related to poor quality of water [18].

17.2 Water-Intensive Agriculture

Indian agriculture consumes 83% of the annually utilised water, whilst the world's average is 69% only. The acute food crisis in the early 1960s prompted the Government of India to introduce high-yielding techniques in agriculture. Substantial increase in food production was achieved, especially that of wheat which was popularly described as the 'green revolution'. But the grass on the other side was not so green. The modified high-yielding seeds were water intensive, demanding several times more water than the traditional variety of seeds. The expansion of irrigation and application of chemical fertiliser and insecticide also played important roles in the increase of food production from 51 million tonnes in 1950–1951 to 209 million tonnes in 1999–2000. The production declined to 196 million tonnes in 2000–2001 [7]. The target of the government is to raise production of food grain around 350 million tonnes by the year 2025 AD. The recent decline in production might be due to severe drought in extensive parts of the country. The irrigated area expanded from 19.50 million hectares to 95 million hectares during preceding five decades of the plan period. The National Water Development Plan is optimistic to provide additional irrigation benefits of 35 million hectares, i.e. 25 million hectares from surface water and 10 million hectares by increased use of groundwater over and above the ultimate irrigation potential of 140 million hectares from all projects. The arithmetic expansion of irrigated area does ensure increase in yield as the yield of cereals per hectare in this country is far below the international standard. It seems to be a paradox that whilst the official godowns are glutted with 62 million tonnes of food grains, about 250 million people below poverty line go hungry everyday! The lack of a good grain distribution system thus affects the food security of the country. The government, however, does not look concerned for the much-needed equity of distribution. The irony is that even with a larger population to feed from a much less arable land compared to that of India, China manages to produce 13% more food grains per capita [4]. Our country also shows little promise of increasing food grain productivity by mid of the twenty-first century. Whilst in experimental farms yield per hectare has already touched 6000 kg per ha, yield levels of only 4000 kg per hectare in irrigated land have been taken as the basis for making projection for food crop production in 2050 [11].

The demand of water in the agricultural sector also continues to increase phenomenally. West Bengal, which was described as having areas of excess water in the colonial documents, now suffers from shortage of water, especially during lean months. The situation has deteriorated after the introduction of green revolution technology in agriculture. The supply–demand deficit presently stands at about 38%. The production of 1 kilogram of dry variety paddy demands 3800 litres of water. The cultivation of such high-yielding paddy in 1 hectare of land consumes so much water that can quench the thirst of 6700 people of a village for the whole of the year. The boro or dry variety paddy is cultivated in West Bengal in about 1,455,000 hectares of land, and the annual production is reported to be 4,415,000 tonnes. This system of production annually consumes 1.78 million ha.m. of water that is more than the total utilisable groundwater resource of the state as a whole [15].

It was this water-guzzling agriculture that brought a total change in the water management scenario in the Indian agriculture [17]. The construction of high dams and reservoirs was once thought to be the solution to the problem. But the large dams were proved inadequate to meet the demand of the water-intensive crops. The dams that were once thought to be temples of development gradually lost its sanctity in the minds of a large number of people. The drought-prone areas of the country are traditionally involved in dry farming and pastoralism. The introduction of water-intensive agriculture in this dry would impair the existing ecosystem. Even in the wet areas, the high-yielding agriculture has posed many problems. This led to the initiation of a new era that witnessed indiscriminate exploitation of groundwater and the privatisation of this natural resource. In the rural areas, farmers established shallow or deep pumps in the midst of their fields and started exploiting groundwater to sell it to the poor and marginal farmers. There is no official regulation to restrict this assault on the nature. The poor and ignorant farmers were thus compelled to buy water that remained stored beneath their own fields. Excessive withdrawal of groundwater led to the contamination of aquifer with arsenic or fluoride. The ecological cost of such misdirected water management system has never been properly estimated. The desiccation of the Aral Sea and massive ecological damage of western USA are two classic examples highlighting the consequences of modifying the natural hydrological systems. But the former NDA Government of India did not care to learn lessons from these experiences. The wrong application of temperate-country technology of water management was again slated to be adopted in our tropical hydroclimate in the name of interlinking of rivers – a massive project that threatens to belittle all engineering feats in the world by its sheer dimension.

17.2.1 Resurrection of a Dormant Proposal

Dr. K.L. Rao [13] first mooted the idea of interbasin transfer of water about three decades back when he proposed excavation of Ganga–Cauvery and Brahmaputra–Ganga link canals. In 1980, Captain D. Dastur put forward a rather emotional scheme of linking peninsular rivers by a garland canal and also the Himalayan rivers by another canal. He further proposed to connect the Himalayan and peninsular components with two pipelines, one through Delhi and another through Patna. But both the proposals of Rao and Dastur were shelved on technical and financial grounds. The National Water Development Agency established in 1982 had been exploring the feasibility of interbasin transfer of water during the last two decades. The main objective was the transfer of water from Ganga–Brahmaputra basin to the river basins of western, central, and southern India. The new Water Policy adopted in 2002 proposed to divide the country into several water zones with a view to the export of water from 'surplus' to 'deficit' zones. The President of India in his address to the Nation on the eve of Independence Day 2002 urged for the interlinking of rivers to meet the challenge of combating the twin menaces of flood and drought. It is important to keep in mind that 2002 was the year of drought. The total monsoon rainfall was about 19% less than the average, and the rainfall in July was 49% less than the preceding years [16]. It was with reference to the Presidential address that a writ petition was filed in the apex court, and the division bench headed by the then Chief Justice gave the landmark judgement on October 31, 2002, suggesting the Central Government to complete the networking of rivers within a decade. The court opined that 'the project will not only give relief to the drought prone areas but will also be an effective flood control measure and would be a form of water harvesting which is being rightly propagated by the Union of India and all the States.' However, date of completion of the project was subsequently extended to the end of 2016. The NDA Government accordingly formed a Task Force under the Chairmanship of Suresh Prabhu to do the needful. The Task Force was asked to explore:

1. All possible means to comply the extremely rigid time schedule directed by the Supreme Court.
2. Management of huge capital that would be required for the execution of the plan.
3. To achieve the consensus amongst States and neighbouring countries.

However, the project is ambitious to achieve following four objectives:

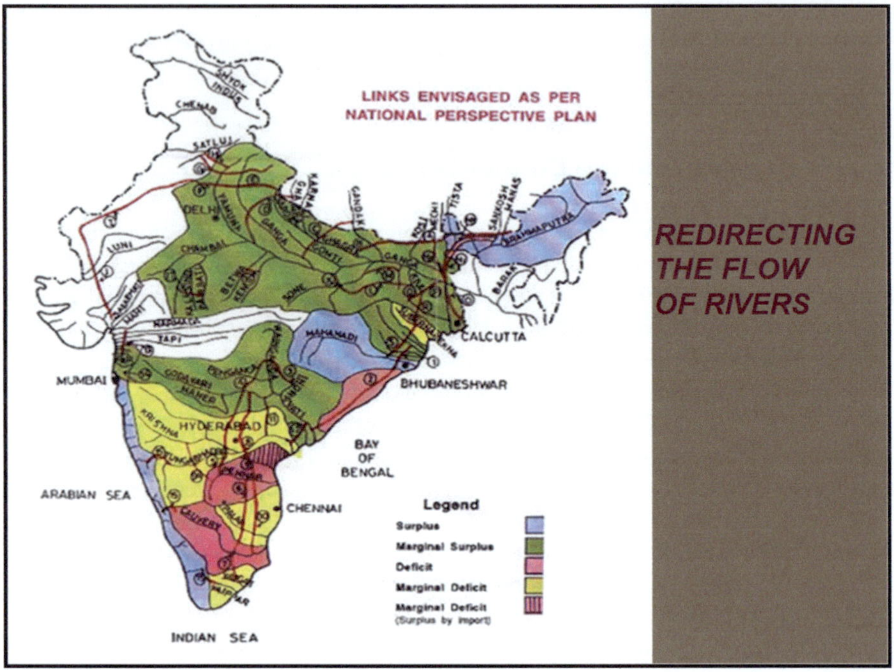

Fig. 17.2 The proposed water zones and link canals

I. To augment the irrigation potential by 35 million hectares (140 Mha to 175 Mha) which would increase the production of food for our growing population
II. To fulfil the growing need of water for the people(the domestic and industrial needs of water is expected to grow steeply by 300–400% in coming decades)
III. To generate about 34,000 MW of additional power through hydroelectricity to meet the increasing need of energy in the country
IV. To facilitate transportation in the country through inland waterways (Fig 17.2)

17.2.2 The Project at a Critical Glance

The National Water Policy (2002) was adopted to ensure food security keeping in view the ever-increasing population that is expected to reach a level of around 139 crore by 2025 AD. The Govt. of India has declared that the production of food grain will have to be raised to about 350 million tonnes to feed this additional population. But this seems to be a gross overestimation because the present population of the country being 103 crore (2001) is being fed with the annual production 208 million tonnes (1999–2000). So the additional food grain requirement of 142 million tonnes by the year 2025 for the 36 crore of additional population cannot be justified.

But the Ministry of Water Resource depends on this inflated data and make the future plan water management.

- The availability of water resource in India is spatially uneven and temporally skewed. The project is optimistic to modify the spatial and temporal inequality of the available water resource within the country and proposes to introduce a controlled water management system. It denies the basic ecological principle that no component in nature is 'excess' or 'wasted' and every drop of rain water plays vital role in the ecosystem.
- The 174 billion cu.m. of water would be diverted from the so-called excess areas to the deficit areas to irrigate 35 million hectares of land. This is 25.22% of the utilisable surface water of the country or more than 98% of the combined live storage capacity of all reservoirs in India. The 14 canals with 9 major dams are supposed to connect the Himalayan rivers, whilst 27 dams and 30 link canals will regulate the flow of peninsular rivers. But dams especially those in Himalaya may induce earthquake as the area is tectonically unstable.
- The plan is also to quench the thirst of people living in 101 districts and five metropolitan cities.
- The hydropower generation target is 34 million KW. But it does not take into account the power that would be required to lift water across the topographic barriers. The power required for such lifting may be greater than that which is to be generated.
- The project also has the 'pious intention' to create 37 million man-years of employment and increase the GDP growth by 4%.
- The estimated cost of this countrywide project is Rs. 560,000 crores. This is twice the revenue that the Government of India currently earns in a year or is about ten times of the total expenditure that the Government has incurred for expansion of irrigation during post-independence era. If we take into account the usual cost escalation, the ultimate cost of the project would exceed the GDP of the country, and the financial management would become an impasse.
- The ecological impacts of the project have not been properly assessed so far. It is said that 79,292 hectares of prime forest may be submerged under the proposed reservoirs of the scheme. But this seems to be a gross underestimation as a single project like Sardar Sarovar is going to destroy 14,000 hectares of forest. We are afraid that a large number of people are going to be evicted for the sake of so-called development though the official document holds the number at 4.50 lakh only.
- The efficiency of the existing dam–canal network is around 35% only. The water-holding capacity of all reservoirs is diminishing at a pace faster than the anticipated rate. The areas affected by waterlogging and salinisation till 1991 were reportedly 2.46 million hectares and 3.30 million hectares, respectively [11]. The problem would be aggravated further with the excavation network of new canals, and the natural drainage system would be largely interrupted.
- Since the excavation of a canal having a discharging capacity of water more than 1400 cumecs (50,000 cusecs.) is hardly possible on technical grounds, the

interlinking can't be a solution to the flood. The peak discharge of the Ganga at Farakka and Brahmaputra at Pandu were measured to be 75,900 cumecs (in September 1998) and 72,726 cumecs (in August 1962), respectively [6, 14]. So withdrawal of 1400 cumecs of water would not moderate the flood in the lower reach. So the concerned water resource experts do not find any logic to justify the proposed interlinking of rivers [19].

- The drought management can be done locally, not by importing water from far ends. There is no village in this country that cannot meet its basic drinking and cooking needs through rainwater harvesting. The keywords should be 'catch the catchment' as was advised by the former Prime Minister in the meeting of the National Water Resource Council on the April 1, 2002. It is not only the scanty rainfall that causes the water crisis but also the technological failure to conserve water. The Cherrapunji and Alwar are two classic examples of 'scarcity in abundance' and 'solution in scarcity', respectively. Instead of opting out for the highly profligate way of supplying domestic water needs which involves centralised collection at faraway points and pipeline distribution systems, as prescribed in the interlinking project, harvesting and community participation in extensive and intensive local water harvesting or 'making water everybody's business' need to be promoted as the only solution to drought [2].
- The Governments of Assam, Kerala and Bihar have already declined to cooperate in the project. The matter relating to water supplies, irrigation and canals, drainage and embankments, water storage and waterpower have been included in the State list (i.e. List-II) of the Indian Constitution. But this authority of the State Government is subject to power of the Central Government mentioned in the entry 56 of List-I that authorised the Union to regulate and control interstate rivers in the public interest. The Parliament has the power to legislate on the development and control of the interstate river. Thus, the Constitution authorises the Union to undertake any project like interlinking of rivers without any consent of the States. But it would be difficult to achieve hydro-solidarity amongst the States when the project would be commissioned. A series of disputes over water like that between Karnataka and Tamilnadu and Punjab and Haryana would emerge leading to strife and discontent between and amongst States.
- The international rules relating to management of transboundary river do not permit unilateral withdrawal of water by a country without any regard to the ecology and environment of the neighbouring country. So Bangladesh has enough ground of resentment.

17.2.3 Changing the Hydraulic Regime

The fluvial system, be it global or local, is an integral part of the hydrological cycle. The river basin is a dynamic system, which continuously transfers water and sediment in the downstream direction. It maintains a dynamic equilibrium in respect of its basin area, number of tributaries, slope, sediment load and discharge. The geometry of the meander is proportional to the discharge flowing along the river. The wavelength and amplitude of meandering pattern has been changing with the variable discharge. The equilibrium of the river is so delicate that any intervention impairs the dynamic functioning of the whole system.

The proponent of the interlinking of rivers often draws the analogy from the linking of national highways, which are static man-made structures on the surface, having no semblance with a dynamic natural system like the river. As connecting one artery with the other in the human body may cause the death of the person concerned, connecting one river basin with the other is likely to invite ecological disaster.

The unmanageable problem relating to such a project would be the crucial sediment management. The Ganga–Brahmaputra system carries more than 1670 million tonnes of sediment annually [8]. Since the construction of the Farakka Barrage in 1971, the mighty Ganga, being impounded, has dropped a substantial portion of sediment load upstream of the barrage. The river has swept away many villages from the left bank and has formed a mighty bend. Now the flow of the river is no longer coaxial to the barrage, and the bays along the northern side have virtually become defunct. Even the possibility of outflanking the barrage by the river cannot be ruled out [14]. The proponents of the interlinking of rivers are reluctant to learn from this experience. They have proposed to interlink all major rivers of West Bengal, and the main objective is diverting water towards South India. The proposed canals are (1) Koshi–Mechi, (2) Brahmaputra–Ganga, (3) Ganga–Damodar–Subarnarekha and (4) Ganga–Sundarbans. The Koshi–Mechi canal is designed to mitigate the flood problem of North Bihar. But the Mechi River is so clogged with boulders and pebbles that it would not be able to accommodate any additional discharge. The induced discharge is likely to aggravate the problem of flood in Mechi basin. It would be more delicate to link Brahmaputra and Ganga. The proposed Brahmaputra–Ganga link through the North Bengal Plains would require interception of no less than 50 south-flowing rivers and that would require construction of a series of aqueducts. Such massive interventions into the river regime would impair the delicate hydrological balance of North Bengal. The execution of the project would require consensus with the neighbouring countries especially Bhutan and Bangladesh. The project is bound to have many impacts on the lower Ganga–Brahmaputra distributaries, and Bangladesh has accordingly already expressed its anxiety over the project in the 35th Meeting of the Indo-Bangladesh River Valley Commission recently held in New Delhi. The international rule does not permit unilateral withdrawal of water from a river that is shared by two countries. The sharing of water must be proportional to the areas occupied by respective countries and population living thereon. The use of water should be rational, equitable and beneficial. The rule does not approve any project that may impair the environment of the neighbouring country.

The proposal to augment the lean months' flow at Farakka was explored in the Indo-Bangladesh agreement in 1996. It was decided to link Sankosh with the Ganga. The 141 km-long Sankosh–Tista link and extension of the same through already existing 26 km-long Teesta–Mahananda link canal would intercept 46 south-flowing rivers en route. The 22 tea gardens covering 530 hectares of land in addition to 770 hectare of prime forest might be destroyed during the excavation of canal. Moreover, the flow along this canal would be against the gravitational pull, and water would be required to be lifted by about 24 to 33 metres. But the canal has not yet been excavated. Now a longer canal named Manas–Sankosh–Teesta–Ganga (MSTG), running almost along the route of the former canal, will be excavated under the interlinking project. We are afraid that the impacts would be equally damaging. The Ganga–Sundarbans and the Ganga–Damodar–Subarnarekha would have a common route for about 300 km. along the west bank of the Bhagirathi. One branch would be extended southwestward towards the Subarnarekha and intercept all east-flowing tributaries to the Bhagirathi. The other branch would flow straight southward and ultimately discharge into the Bay of Bengal. The Ganga–Sundarbans canal is a misnomer as the famous mangrove forest lies on the eastern littoral tract whilst the canal is proposed along the west bank of the Bhagirathi. The other branch of the canal would have to negotiate the undulating Rarh plateau of West Bengal whilst linking the Damodar with Subarnarekha.

In the deltaic tract, rivers frequently change their courses. This is a natural process that governs sediment and nutrient distribution during the floods. The existing dams and barrages of the Ganga basin have to a great extent intercepted the process of sediment transfer. The proposed interlinking would further interrupt this natural distribution system and thereby impair the existing ecosystem. The rivers would significantly alter the present geometry of meanders, and riparian settlements are likely to be threatened by the gnawing rivers.

17.2.4 Myth of Excess Water

In contravention to the law of nature that there is nothing excess in its perfectly balanced systems, the Government of India has decided to divert "excess water" from the Ganga–Brahmaputra basin to the arid west and parched South India. The Ganges–Brahmaputra–Barak basin is endowed with more than 62% of the nation's water resource, but its temporal distribution is extremely skewed. More than 80% total annual discharge flows during the 3 monsoon months of July–September. The Ganga basin drains about 26% of the Indian landmass, and water resource potential of the basin is 525 km^3. The Ganga basin in India is currently the home of 356.8 million populations. The average density of 413/km^2 compared to about 256.3/km^2 for the whole of India indicates a very high level of demand generation on the basin water resources [11]. The per capita available water is about 1473 m^3, and this is less than the national average (1820 km^3). Still the Government is planning to divert water from the Ganga basin. The Brahmaputra basin in India covers only 7.7% of the national territory and enriched with 537.3 km^3 of water resource. But per capita available water in this part of the nation is strikingly as high as 18,417 m^3. But it is not denied in the official record of the Government of India that out of 690 km^3 of utilisable surface water resource, the Ganga–Brahmaputra basin carries only 40% whilst the share of all other basins is 60%. The Ganga basin being the home of about 700 million people has been suffering from acute water stress during lean months (Fig. 17.3).

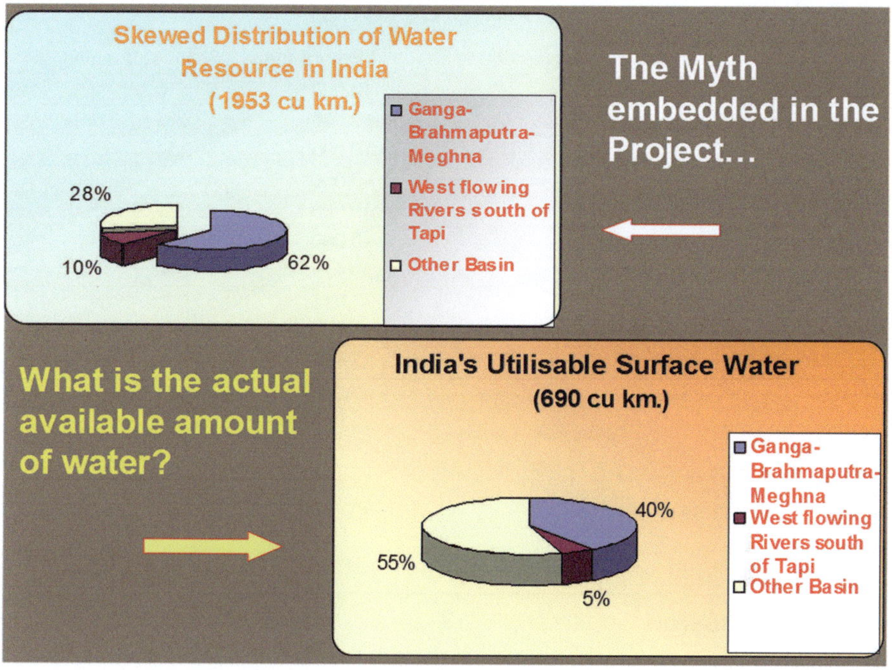

Fig. 17.3 Myth of excess water

It cannot be ignored that Bangladesh too depends largely on the Brahmaputra water resource. Consideration of the population of Bangladesh would further drastically reduce the per capita availability of water in the Brahmaputra basin. The basin is endowed with 41% of India's hydropower potential, and only 3% of it has so far been exploited [6]. The underexploited water and power resource of the Brahmaputra basin now appears to be very lucrative to the Planners of India. But any withdrawal of water from GBB needs prior consensus with Bangladesh. The issue of sharing Ganga water has always been delicate relating to Indo-Bangladesh relationship. The lean months' flow at Farakka remains far below the threshold limit required for meeting the demand of both the countries (Table 17.3).

Bangladesh has already rejected the proposal of the excavation of Brahmaputra–Ganga link canal through its territory to augment the flow at Farakka. Now the second alternative of aligning the link canal through the North Bengal plain is being explored. The Government without any regard to ecology and international relationship has accepted this proposal.

Table 17.3 Indo-Bangladesh sharing of the Ganga waters 1996

Time	Avg. flow at Farakka (1949–1988)	Share of India	Share of Bangladesh
Jan 1–10	107,516	40,000	67,516
11–20	97,673	40,000	57,673
21–31	90,154	40,000	50,154
Feb 1–10	86,323	40,000	46,323
11–20	82,859	40,000	42,859
21–28	79,106	40,000	39,106
Mar 1–10	74,419	39,419	35,000
11–20	68,971	33,931	35,000
21–31	54,688	35,000	29,688
Apr 1–10	63,180	28,180	35,000
11–20	62,633	35,000	27,633
21–30	60,992	25,990	35,000
May 1–10	67,351	35,000	32,351
11–20	73,590	38,590	35,000
21–31	81,854	40,000	41,854

Figures in Cusecs

17.3 Conclusion

This seems to be the declaration of crusade against the nature, and many scientists believe that the proponents of the project are going to invite an ecological disaster that may mortgage the generations to come. The idea of excess water in any river basin is an unmixed myth because the peak discharge flushes the sediment load that is deposited in the distributaries during lean months. It is further believed by the modern river scientists that flood is not necessarily the evil but is a natural hydrological event that restores the ecological balance of the delta. The modern theory of flood management does not address structural measures but rather emphasises on community-based disaster preparedness. The CBDP includes early warning, identification of flood shelter, ensuring the drinking water during flood, conservation of food and medicine, changing crop calendar in low-lying areas and finally preparedness for rescue–relief–rehabilitation.

The freshets of the monsoon in the Ganga–Brahmaputra delta perform the twin ecological functions of recharging the groundwater and combating the saline intrusion from the sea. The sweet and brackish water regime of the delta is so juxtaposed that such a massive intervention in the fluvial regime would impair the entire ecosystem. The big question is that whether water can really be transferred to the parched areas? The existing canal systems in the lateritic tract of West Bengal absorbs as high as 66% of water, and the tail end of the command area hardly receives any irrigation water during lean months when demand is at its peak. Had this been the case of West Bengal where soil and air carry much moisture, the loss of water in parched area would naturally be much higher. In view of monumental cost of the project to the tune of Rs. 560,000 crores and inevitable cost escalation, the water, if reaches the tail end at all, would remain beyond the affordability of resource-poor farmers. The National Water Policy (2002) clearly stated that 'the water charges for various uses should be fixed in such a way that they cover at least the operation and maintenance charges of providing the service initially and a part of capital costs subsequently'. The former Prime Minister himself admitted in his speech at the meeting of the National Water Resource Council that the 'community is the rightful custodian of water'. He further added 'Technologies and methods are available today whereby the agriculture sector could cut its water needs by ten to fifty percent, industries by forty to ninety percent and cities by thirty to thirty-five percent without any sacrifice of economic output or quality of life'. But Suresh Prabhu, the former Chairman of the task force, did not care about the earlier advice of the Prime Minister and was keen to take up the project of interlinking of rivers on a war footing. The President of India and the apex court were not fractious on this issue. The stage was set to convert the natural resource to a commodity.

References

1. Agarwal A, Narain S (1991) Floods, flood plains and environmental myths. CSE, New Delhi. 167p
2. Agarwal A, Narain S, Khurana I (2001) Making water everybody's business. CSE, New Delhi. 456p
3. Bandyopadhyay, Mallik J, Mandal BM, Perveen S (2002) Report on the policy dialogue on dams and development. EPW XXXVIII(40):4108–4112
4. Bandyopadhyay J, Perveen S (2003) The interlinking of Indian Rivers- some questions on the scientific, Economic and environmental dimensions of the proposal, Paper presented at Seminar on interlinking Indian Rivers: Bane or Boon? at IISWBM, Kolkata, ACB Publications, Kolkata
5. Worster D (1985) Rivers of empire. OUP, New York, p 20
6. Goswami DC, Das PJ (2003) The Brahmaputra River in India. Ecol Asia 11(1):9–14. Mumbai
7. Hindu (2002) Survey of Indian Agriculture, p 5
8. Milliman JD, Meade RH (1983) World-wide delivery of River sediment to the Oceans. J Geol 91(1):1–21
9. Ministry of Water Resources (2003) 4th editors conference on social sector issues, pp 3–4
10. Mishra DK (2002) Living with the politics of floods. Peoples' Science Institute, Dehradun. 124p
11. National Commission for Integrated Water Resource Development Plan (NCIWRDP) (1999) Integrated water resource development: a plan for action september, Ministry of Water Resources, and New Delhi, pp 1–97
12. Govt. of India (2003) New Delhi, India, pp 435–455
13. Rao KL (1975) India's water wealth. Orient Longman, New Delhi, p 231
14. Rudra K (2000) Living on the edge: the experience along the Bank of Ganga in Malda District, West Bengal. Ind J Geog Environ 5:57–67
15. Rudra K (2003) Paschimbanger Krishiniti – Krishaker Bhabisyat (in Bengali), a collection of articles on agricultural policy of West Bengal. 7(16):138–149
16. SANDRAP (2003, February) Dams, Rivers and People New Delhi, p 37
17. Shiva V (2002) Water wars: privatisation, pollution and profit. IRP, New Delhi, p p159
18. Shiva V, Bhar RH, Jafri AH, Jalees K (2002) Corporate Hijack of water. Navadanya, New Delhi, p 3
19. Singh B (2003) A big dream of little logic. Hindustan Times March 09, 2003, Kolkata p11

Dr. Kalyan Rudra is presently chairman of the West Bengal Pollution Control Board (WBPCB). He is a geographer by academic training having specialization in river and water management. He took a break from his protracted teaching carrier in 2007 and joined in a research project of the West Bengal Pollution Control Board. He worked as the project director in two projects of WBPCB – one on water resource and its quality and the other on the dynamic river system of West Bengal. He has been the expert member in the committee constituted by the apex court of the state for cleaning the Ganga since 2005. Dr. Rudra was a member of the National Flood Management Core Group and headed the committee constituted to advise the Government of West Bengal on the issue of Indo-Bangladesh sharing of Teesta water. He is regular guest faculty in the National Academy of Audit and Accounts and Academic Staff Colleges of Calcutta and Burdwan University. He has written 5 books and 30 research papers which have been published in reputed journals of India and abroad. Dr. Rudra was also advisor to the International Union for Conservation of Nature. Presently, he has been working on the history of mapping Bengal.

Cost-Benefit Analysis of Hydropower: Case Study of Kotlibhel 1B

18

Bharat Jhunjhunwala

18.1 Introduction

Government of Uttarakhand is making a cascade of run-of-river (ROR) hydropower dams on all its rivers. The idea is that this will provide immense economic benefits to the people of the state. Therefore, some environmental costs have to be borne, howsoever reluctantly. This principle has been enunciated by the Hon'ble Supreme Court in various judgements in terms of 'balancing the right to development with the right to life'. The author believes this logic is correct.

The simple method for doing this is cost-benefit analysis (CBA). This principle is also enshrined in the National Relief and Rehabilitation Policy 2007 which states that costs and benefits will be transparently placed before the public.

The question raised in this chapter is whether the stated economic benefits of the ROR hydropower projects are greater than the costs.

As a rule, however, hydropower projects and the Central Electricity Authority does not make a CBA. The Expert Appraisal Committee for river valley projects of the Ministry of Environment and Forests (MOEF) also does not require a CBA to be presented. There is no requirement to place a money value on the environmental impacts of the projects. However, project authorities are statutorily required to file a cost-benefit analysis with Ministry of Environment and Forests in the application for diversion of forest land under Forest Conservation Act (FCA).

The author has studied the CBA filed under FCA for a few ROR projects. A case study of the CBA for Kotlibhel 1B (KB1B) project proposed by NHPC [1] on river Alaknanda upstream Devprayag is given in this chapter as a case study. The CBA filed by NHPC is attached as Annexures 1 and 2 of this paper along with the author's comments.

B. Jhunjhunwala (✉)
Lakshmoli, PO Maletha, Via Kirti Nagar, Tehri 249161, Uttarakhand, India
e-mail: bharatjj@gmail.com

K.J. Nath, V.P. Sharma (eds.), *Water and Sanitation in the New Millennium*,
DOI 10.1007/978-81-322-3745-7_18

18.2 Benefits and Costs

18.2.1 Benefit 1: Electricity

The main benefit from a ROR project is generation of electricity. NHPC has given the following statements:

- Increase in productivity attributable to the specific project is shown as (+) Rs 300.6 crores per year (1268 MU energy at 2.37 per unit) (Annexure 2, line1). This is actually the cost of production of energy. This is not increase in productivity as shown by NHPC.
- Actually, the increase in productivity is the difference between the cost of production of electricity by the project and the cost of production from alternative sources. If the power produced from KB1B costs Rs 2.37 per unit while the next cheapest source of power is, say, Rs 5.00 per unit, then the benefit to the society from KB1B will be Rs 2.63 per unit (5.00–2.37). This is the money the society will save because of implementation of the project. The society will have to pay Rs 5.00 for the electricity in absence of the project and only Rs 2.37 with the implementation of the project. The difference is the 'increase in productivity attributable to the specific project'.

The author assesses this at Rs 3.30 per unit on the following basis:

- NTPC Vidyut Vyapar tells us the sale price of peak power was Rs 2.65 against average price of Rs 2.05 in 2004 [2]. The difference between price of peak power and average price is Rs 0.60 per unit
- The average sale price of peak power in 2004 by NTPC Vidyut Vyapar was Rs 2.65 per unit. This is a good proxy of marginal product. This works out to Rs 2.87 per unit at 2006 prices. Against this, Rs 3.30 per unit has been assumed. The sale price of Rs 5 to Rs 6 for peak power, as reported from anecdotal sources, is probably the price for short-term demand in special circumstances.
- The sale price of base power is Rs 2. The maximum sale price of peak power is Rs 5 to Rs 6. The true price is likely to be somewhere in the middle.
- NHPC appears to have assumed a price of Rs 2.00 for base power and Rs 4.00 for peak power leading to average of Rs 2.80 as mentioned in DPR-ES.

On these considerations, a value of Rs 3.30 per unit for marginal product seems appropriate. This marginal product is for the average of peak and nonpeak power. The benefits from Kotlibhel 1B are calculated in Table 18.1, based on above value of power.

It may be noted that the price at which power is sold by NHPC makes no difference to this calculation. Increase in sale price raises the benefits to NHPC and reduces the benefits to consumer. Reduction in sale price does the opposite. The total remains unchanged in either case.

Table 18.1 Benefits of electricity generated from Kotlibhel 1B

Sl no	Item	Amount	Basis
1	Total sale of electric power	111.6 crore units	DPR-ES [1]. This excludes free power to State, benefits from which are calculated separately in the next section
2	Cost of production	Rs 2.37 per unit	DPR-ES
3	Marginal product	Rs 3.30 per unit	Author's estimate as above
4	Social benefit per unit	Rs 0.93 per unit	Line 3 – line 2
4	Social benefit total	Rs 103.8 crore per year	111.6 crore units × Rs 0.93 per unit

Table 18.2 Benefits of free power to the State

Sl No	Item	Value	Basis
1	Free peak power to state	15.2 crore units	Table above
2	Marginal product or total benefit	Rs 3.30 per unit	Cost is assumed as zero since this is supplied free.
3	Total benefit to state	Rs 50.2 crore per year	

It must be stressed that the above calculation takes the profits of NHPC to be synonymous with social benefits. This is done because NHPC is a public sector unit. The profits belong to the nation. The calculation would be different for private players. In that case, the sale price of power would be deducted from social benefits, instead of cost of production as done here.

18.2.2 Benefit 2: Free Power to Host State

NHPC is to supply 12% free power to the host state. The benefits to Uttarakhand from this are estimated in Table 18.2.

18.2.3 Benefit 3: Employment

NHPC has stated that 1200 jobs will be created (Annexure 2, line 4) of NHPC. The question is how much 'benefit' does this translate into? NHPC does not give any assessment of the money value of the benefit that could be incorporated in the CBA.

The figure of 1200 appears to be exaggerated. ROR projects do not generate direct employment in such numbers. Perhaps NHPC has included 'indirect' employment video shops, taxis, truck, etc. The author submits that such indirect employment should not be included in benefits because much loss of indirect employment also takes place from submergence of agricultural and forest land, in particular. Hence, to include indirect gain of employment while ignoring indirect loss of

Table 18.3 Benefit from employment due to KB1B

Sl no	Item	Amount	Basis
1	Number of direct employees	150 numbers	Author's best estimate
2	Average wages of NHPC employees	Rs 1.33 lacs per year per employee	NHPC Balance Sheet 2005–2006 gives number of employees at 13,118 at page 6. Sales are Rs 1614 crores of which 10.84% are employee's remuneration (page 23)
3	Opportunity cost or income lost in alternative employment by local employees	(−) Rs 36,000 per year per employees	The 'benefit' from Kotlibhel 1B would be equal to the total salary paid, less incomes obtained by the same persons in absence of Kotlibhel 1B. The author's best estimate of this opportunity income is Rs 36,000 per year
4	Net income added due to Kotlibhel 1B HEP	Rs 0.97 lacs per year per employee	Line 1 – line 2
5	Benefits or the additional wages paid by KB1B to local employees	Rs 145.5 lacs per year or Rs 22.8 crore on lifetime of project	For 150 employees

employment is fallacious. Therefore, the author relies on his best estimate that direct employment of 150 persons will be generated from KB1B. The benefit from this employment is given in Table 18.3.

18.2.4 Other Costs

NHPC has given 'particulars for evaluation of loss of forest' which is annexed as Annexure 1 of this chapter. This statement suffers from many infirmities that are shown at Annexure 1.

More importantly, NHPC ignores many costs. These have been calculated by the author [3]. A summary of these costs is given at Table 18.4.

The author recognizes that many of these estimates may be challenged and alternatives be suggested. The author welcomes such intervention. The point is that all these costs are wholly ignored in the CBA filed by NHPC.

18.2.5 Cost-Benefit Analysis

A perusal of above tables gives us a CBA which is depicted in Table 18.5.

Table 18.4 Environmental costs on which NHPC is silent in its calculations

Sl no	Item	Cost (Rupees Crore/Year)
1	Sediment	98.0
2	Quality of water	350.0
3	Methane emissions	62.8
4	Forests	61.1
5	Earthquakes	8.4
6	Landslides	2.9
7	Malaria and health	6.4
8	Biodiversity	11.7
9	Otters	20.0
10	Road accidents	7.1
11	Decline in temperatures	7.0
12	Sand	18.2
13	River rafting	8.0
14	Bridges	4.9
15	Aesthetic value of free-flowing water	60.5
16	Immersion of ashes	5.4
17	Relocation of temples	4.2
18	Loss of fishing	2.5
19	Cascade effect	192.7
20	Total cost of KB1B	931.8

Table 18.5 Cost-benefit analysis (CBA) of KB1B

Sl No	Item	Benefit	Cost
1	Benefits from generation of power	103.8	
2	12% Free power to State	50.2	
3	Employment	1.5	
4	Costs as at Table 18.4		931.8
	Total benefit and cost of Kotlibhel 1B HEP	155.5	931.8
	Net loss		776.3

18.2.6 Political Economy of Hydropower

Why are such projects made when costs are larger than benefits is a puzzle. This mystery is unravelled when one examines the distribution of benefits and costs between different stakeholders.

The share of salaries and wages in the total turnover of NHPC are much higher than private hydropower companies. This is clear from comparison of Balance Sheet of NHPC and Jaiprakash Hydropower (Table 18.6).

Salaries and wages constitute 12.5% of NHPC's turnover against 1.8% for Jaiprakash Hydropower. NHPC has contented that the figures for a small company like Jaiprakash Hydropower are not comparable to those of a large company like NHPC; hence, higher share of salaries and wages is justified. Indeed this much is true that the figures are not exactly comparable. But the relationship is exactly

Table 18.6 Share of salaries and wages in cost of power, 2006–2007

Sl No	Item	Jaiprakash Hydropower [4]	NHPC (2007)
1	Total sales	Rs 335.8 crore	Rs 1882.9 crore
2	Salaries and wages (amount)	Rs 5.9 crore	Rs 236.0 crore
3	Salaries and wages (percent)	1.8%	12.5%

Table 18.7 Share of salaries in state government expenditures

Sl no	State (Year)	Salary	Pension	Study link
1	Kerala (2004–2005)	31%	15%	Kerala Public Expenditure Review Committee, First Report, May 2006 [5]
2	Tamil Nadu (2000–2001)	39%	13%	White Paper On Tamil Nadu Government's Finances [6]
3	All India (1999–2001)	32.1%	NA	MOF [7]
	Average	34.0%	14%	

opposite of the claim by NHPC. The share of employee cost should be reduced in a large company because it gains from economies of scale. Thus, the share should be lower rather than higher.

Be that as it may, the main objective is to unravel the mystery of why such loss-proposition projects are imposed upon the country. On pro rata basis, a sale of Rs 973.1 crores projected from Kotlibhel 1B will provide for salaries of Rs 121.6 crores for the employees of NHPC [1]. This is the gain to all employees of the company, including head office, CMD office, etc.

(*This figure should not be confused with the benefits from additional employment generated in Kotlibhel 1B, which take into account only employees at the field level.)

The share of employees of Government of Uttarakhand in the free power to be supplied by NHPC from Kotlibhel is similarly assessed. The author has been able to access three studies which give figures of share of state income appropriated by the employees (Table 18.7).

Based on these studies, one may assume that 48% of the revenues of the Government of Uttarakhand will be spent towards salaries and pensions of the government employees. Thus, out of Rs 50.2 crore received by Government of Uttarakhand, the State Government employees may get an additional Rs 24.1 crores, and the people may get Rs 26.1 crores.

Local people get huge compensation for the land. This cost is built into the cost of production of power. The additional benefit to local people is assumed to be equal to the R&R package of Rs 13.2 crores made by NHPC. This is one-time cost; hence, the annualized benefit is taken at 10% of Rs 1.3 crores per year. Now one can assess the distribution of benefits and costs from Kotlibhel 1B (Table 18.8).

Table 18.8 Distribution of benefits (−) and costs (+) (rupees crore per year)

Chapter no	Item	Total	Employees of NHPC and Government of Uttarakhand	People of India
1	Benefits from generation of power	(+) 103.8	–	(+) 103.8
2	12% Free power to State	(+) 50.2	(+) 24.1	(+) 26.1
3	Employment	(+) 1.5	–	(+) 1.5
4	Costs	(−) 931.8	–	(−) 931.8
	Total	(−) 776.3	(+) 24.1	(−) 800.4
5	Memo: benefits to employees of NHPC in generation of electricity	–	(+) 121.6	–
6	Memo: compensation for land	–		(+) 1.3
7	Total, including memo items		(+) 145.7	(−) 799.1

The above chart gives an indication of the gainers and losers.

Gainers: NHPC and state government employees to the tune of Rs 145.7 crores.
Losers: People of the country to the tune of Rs 799.1 crores.

Kotlibhel 1B HEP is a loss proposition for the nation. Yet it is being implemented because it imposes hidden costs on the people and provides direct benefits to employees of NHPC and government of Uttarakhand.

A possibility is that KB1B is beneficial for the people of Uttarakhand even if it is harmful for the people of India. The objective of GOUK is to secure benefit of people of Uttarakhand.

In order to assess this, it is instructive to break up the benefits and costs by stakeholders: (1) employees of NHPC and GOUK, (2) affected people, (3) people of Uttarakhand and (4) people of India. This is done by distributing the costs and benefits as shown in Table 18.9 between the four stakeholders as per best estimates of the author. The reader is welcome to make assessment of the distribution, and the same would be accepted by this author with respect. The main point is of methodology. It is imperative that one looks at the distribution of costs and benefits over different stakeholders.

The breakup of costs and benefits is as follows:

Gainers: NHPC and state government employees to the tune of Rs 145.7 crores.

Losers 1: Affected people to the tune of Rs 81.1 crores.
Losers 2: People of Uttarakhand to the tune of Rs 71.4 crores.
Losers 3: People of India to the tune of Rs 646.6 crores.

The author welcomes alternative analysis of costs, benefits and their distribution and the same being put to public scrutiny. But it is unacceptable not to undertake such stakeholder analysis.

Table 18.9 Distribution of costs and benefits of KB1B by stakeholders (Rs crore/year)

Sl no	Item	Total	Ratio of distribution	Employees of NHPC and GOUK	Affected people	People of Uttarakhand	People of India
1	Benefits from generation of power	(+) 103.8	0-01-11-88	–	(+) 1.0	(+) 11.4	(+) 91.4
2	12% Free power to State	(+) 50.2	48-01-51-0	(+) 24.1	(+) 0.5	(+) 25.6	–
3	Employment	(+) 1.5	0-33-34-33	–	(+) 0.5	(+) 0.5	(+) 0.5
4	Sediment	(–) 98.0	0-0-0-100	–	–	–	(–) 98.0
5	Quality of water	(–) 350.0	0-1-10-89	–	(–) 3.5	(–) 35.0	(–) 311.5
6	Methane emissions	(–) 62.8	0-0-1-99	–	–	(–) 0.6	(–) 62.2
7	Forests	(–) 61.1	0-25-50-25	–	(–) 15.3	(–) 30.5	(–) 15.3
8	Earthquakes	(–) 8.4	0-75-25-0	–	(–) 6.3	(–) 2.1	–
9	Landslides	(–) 2.9	0-100-0-0	–	(–) 2.9	–	–
10	Malaria and health	(–) 6.4	0-50-25-25	–	(–) 3.2	(–) 1.6	(–) 1.6
11	Biodiversity	(–) 11.7	0-1-1-98	–	(–) 0.1	(–) 0.1	(–) 11.5
12	Otters	(–) 20.0	0-1-1-98	–	(–) 0.2	(–) 0.2	(–) 19.6
13	Road accidents	(–) 7.1	0-25-50-25	–	(–) 1.8	(–) 3.5	(–) 1.8
14	Decline in temperatures	(–) 7.0	0-75-25-0	–	(–) 5.2	(–) 1.8	–
15	Sand	(–) 18.2	0-75-25-0	–	(–) 13.6	(–) 4.6	–
16	River rafting	(–) 8.0	0-50-25-25	–	(–) 4.0	(–) 2.0	(–) 2.0
17	Bridges	(–) 4.9	0-75-25-0	–	(–) 3.7	(–) 1.2	–

(continued)

Table 18.9 (continued)

Sl no	Item	Total	Ratio of distribution	Employees of NHPC and GOUK	Affected people	People of Uttarakhand	People of India
18	Aesthetic value of free-flowing water	(−) 60.5	0-1-1-98	–	(−) 0.6	(−) 0.6	(−) 59.3
19	Immersion of ashes	(−) 5.4	0-25-75-0	–	(−) 1.3	(−) 4.1	–
20	Relocation of temples	(−) 4.2	0-25-75-0	–	(−) 1.0	(−) 3.2	–
21	Loss of fishing	(−) 2.5	0-75-25-0	–	(−) 1.9	(−) 0.6	–
22	Memo: total of costs	(−) 583.6	–	(+) 24.1	(−) 62.6	(−) 54.2	(−) 490.9
23	Cascade effect	(−) 192.7	In ratio as at line 22	–	(−) 19.8	(−) 17.2	(−) 155.7
24	Costs and benefits of KB1B	(−) 776.3	–	(+) 24.1	(−) 82.4	(−) 71.4	(−) 646.6
25	Memo: benefits to employees of NHPC in generation of electricity	(+) 121.6	100-0-0-0	(+) 121.6	–	–	–
26	Memo: compensation for land	(+) 1.3	0-100-0-0	–	(+) 1.3	–	–
27	Final costs and benefits of KB1B	(−) 653.4	–	(+) 145.7	(−) 81.1	(−) 71.4	(−) 646.6

Annexures

Annexure 1: Critique of Costs of KB1B as per NHPC

(After Annexure 2b "Particulars for Evaluation of Loss of Forests" filed by NHPC with MOEF)

Sl no	Parameter	As per company	This author's comment	This author's estimate, one-time unless specified	Basis of this author's estimate
1	Loss of value of timber, fuel wood and minor forest produce on an annual basis including loss of man-hours per annum of people who derive livelihood and wages from the harvest of these commodities	Cost of 4405 trees Rs 43 lacs (one-time); loss of livelihood Rs 4.39 lacs (annual)	Value of Rs 997 per tree is grossly undervalued. True value is about Rs 100,000 per tree. Annual loss of livelihood discounted to NPV should be given	(−) 44 crores	Value of trees at 1 lac.
				(−) 70.2 crores	Value of livelihood at 4.4 crores per year = 70.2 crores at 5% discount rate
2	Loss of animal husbandry productivity, including loss of fodder	50% of loss of annual livelihood = 2.19 lacs (annual)	Assuming 50% of loss of livelihood is arbitrary and without basis	(−) 58.3 crores	NPV is taken as 5.9 lacs per ha. Assume equal amount due to loss of ecotourism, ecotourism and recreation benefits, watershed benefits, microclimatic factors, carbon sink, biodiversity and employment generation through forestry works. Again assume same value for biodiversity. Total Rs 11.8 crores per ha or Rs 58.3 crores discounted value
3	Cost of human resettlement	Nil	It is assumed that *every affected family* has house in neighbouring village and prefer to stay there. This is a false statement made without basis	(−) 25 crore (lump sum)	Displacement due to HEPs affects not only those displaced but also those left untouched by splitting their society. Many studies have detailed the loss of self-identity of the displaced and trauma and alienation of all concerned [8]

4	Loss of public facilities and administrative infrastructure	(−) 33.22 crores	Basis of valuation appears to be Rs 6 lacs per ha. This is arbitrary. Value of new roads is shown as nil. Three bridges on Alaknanda will be submerged	Land (−) 750 crores Bridges (−) 9 crores	Final value is likely to be Rs 1.5 crore per ha since compensation offered by nearby Srinagar HEP in 2007 is Rs 1.0 crore per ha. Value of these bridges is estimated at Rs 2 crore each. Add value of approach roads, electricity lines, water supply pipe lines, etc. Rs 3 crores each
5	Environmental losses (soil erosion, effect on hydrological cycle, wildlife habitat, microclimate upsetting of ecological balance)	(−) 6.28 lacs	Based on value of trees submerged only. This is really double accounting since value of trees is already accounted at line no 1. All other environmental losses are ignored	(−) 931.8 crores per year	Estimated by author in cost of hydropower [3]
6	Suffering to oustees	(−) 7.85 lacs	Appears to be based on income loss of 33 fully displaced families x 2 persons per family x Rs 7937 per capita income x 1.5 years. The number of displaced persons, number of persons per family and per capita income are all misrepresented	(−) 33.3 crores for direct income lost (−) 33.3 crores for social trauma	Number of displaced persons is 550 if we take 25% loss of land as the criterion as done in Sardar Sarovar. The EIA gives incomes of families as follows: 5.4 families x Rs 750; 24.6 families x Rs 3000, 9.6 families x Rs 7500, 2.4 families x Rs 12,500 (page 159). Average is Rs 4283 per month or Rs 51,400 per year per family. The number of families affected is 550. Assume they lose 50% of their income. Multiply by 1.5 times (as per NHPC). Hence, loss of income is 51,400 x 550 x 0.5 x 1.5 = 2.12 crore per year. The NPV for 30 years = 33.3 crores (at 5% discount rate)

Annexure 2: Critique of Benefits of KB1B as per NHPC

(After Annexure 2a "Evaluation of Benefit, Notwithstanding Loss of Forest" filed by NHPC with MOEF)

1	Increase in productivity attributable to the specific project	(+) 300.6 crores per year.	1268 MU energy at 2.37 per unit. This is actually the cost of production of energy. This is not increase in productivity	(+) 298.3 crores during lifetime of project	Increase in productivity is the difference between the cost of production and the cost of energy from alternative sources (Details at Table 18.8)
2	Benefits to economy	No value given	It is claimed that industrial and agricultural production will increase manifold. The gain from electricity is already incorporated in line 1. The loss to agricultural productivity due to lower temperature is not accounted for	No Comment	–
3	No of benefited population (value judgement)	Value judgement	Population of states receiving power will be benefited	–	This benefit is already incorporated in line 1 of this table
4	Employment potential (value judgement)	Value judgement	1200 additional jobs will be available. This includes indirect employment generated. The indirect benefit, however, has to be adjusted against indirect losses of employment and income from submergence of forest land and private land, lower yields from lower temperatures, etc. NHPC errs in including indirect creation of employment while excluding indirect destruction of employment	(+) 22.8 crores	Table 18.2
5	Cost of acquisition of facility on non-forest land wherever feasible	–	NHPC says project is site specific; hence, no other alternative is available	–	Alternatives are available in terms of solar, wind, micro-hydro, demand side management, efficient use of power, etc.

6	Loss of agriculture and animal husbandry production due to diversion of forest land	Nil	No explanation is given	No comment	–
7	Cost of rehabilitation of displaced persons as different from compensatory amounts given for displacement	(–) 13.2 crores	This is *cost* of R&R package as provided by NHPC in the project proposal. This is not the final cost as may be provided by the concerned authorities, which is likely to be many time more	(–) 39.6 crores	Assume the final R&R package will be three times the amount provided for by NHPC in the project proposal
8	Cost of supply of free fuelwood to workers residing in or near forest area during the period of construction	(–) 0.7 crores	Cost of fuel supplied to NHPC workers	(–)0.7 crores	No comment

References

1. NHPC (2005) National hydroelectric power corporation, Kotlibhel Hydroelectric Project Stage-1B (Uttaranchal), Detailed project report—executive summary, Faridabad
2. NTPC (2004) Vidyut Vyapar Nigam Limited, Performa for submission of information for the Quarter (Jan 2004 to March 2004), Form –III, http://nvvn.co.in/form_III_III.html
3. Jhunjhunwala (2009) Bharat, Economics of Hydropower, Kalpaz
4. Rediff (2007) Jai Prakash Hydro-Power Ltd. http://money.rediff.com/money/jsp/p_l.jsp?companyCode=15130060
5. Kerala (2006) Public expenditure review committee, First Report, May 2006. http://www.kerala.gov.in/dept_finance/perc_1rep.pdf
6. TN (2007) White Paper on Tamil Nadu Government's Finances, Government of Tamil Nadu, http://www.tn.gov.in/budget/archives/white5-1.htm
7. MOF (2003) Economic survey 2002–2003, Review of developments, Ministry of Finance, Government of India
8. WCD (2000) Dams and development, World Commission on Dams, Earthscan, London

Dr. Bharat Jhunjhunwala The author acquired his bachelor's degree in science (physics, chemistry and mathematics) from Kanpur University and doctoral degree in food and resource economics from the University of Florida in 1973 at the age of 23.

He was assistant professor of economics at the Indian Institute of Management Bangalore. He was consultant to donors and NGOs like Oxfam, Care, Swedish International Development Cooperation Agency and Swiss Development Cooperation in the 1990s.

He has undertaken extensive studies on economic impacts of hydropower dams. He has published three books on the topic: (1) *Economics of Hydropower*, (2) *Economics of River Flows: Lessons from Dam Removals in America* and (3) *Water: Impact of Dams on its Qualities*.

He has been involved in many court cases relating to the construction of dams on rivers. A study on the cumulative impact assessment of dams on the Ganga River was commissioned by MoEF pursuant to his intervention. Two projects, namely, Kotlibhel 1B and Kotlibhel 2, on the Ganga River have been cancelled or held up because of these litigations. He has argued personally in many cases before the Supreme Court, National Green Tribunal and High Court of Uttarakhand.

Some of his major books on other topics are as follows:

- *Welfare State and Globalization: A Critique of Amartya Sen* (Rawat, Jaipur, India, 2000)
- *Governance and Human Rights*, edited (Kalpaz, New Delhi, 2002).
- *Human Development*, with R.K. Bhattacharya and V.K. Srivastava, co-editors (Kalpaz, New Delhi, 2003)
- *Traditional Agricultural and Water Technologies of the Thar* (Gravis, Jodhpur, India, 2003)

He writes a weekly column on political economy which is published in about 20 papers in 10 languages in India.

He was awarded the Nachiketa Award by Prime Minister Atal Bihari Vajpayee in 2002 for outstanding contribution to economic journalism.

Impact of Water Contamination and Lack of Sanitation and Hygiene on the Nutritional Status of the Communities

19

Indira Chakravarty

19.1 Introduction

The importance of safe water, sanitation and hygiene to maintain good health is a known fact, because these prevent infections of various types which eventually lead to poor health and nutrition. However, in many ways, WASH (water, sanitation and hygiene) also has a direct impact on the nutritional status of communities in a number of ways. For this, maintenance of water quality along with easy availability is essential [1].

19.1.1 Water Quality

Although appearance, taste and odour are useful indicators of the quality of drinking water, suitability in terms of public health is determined by microbiological, physical, chemical and radiological characteristics. Of these, the most important is perhaps microbiological quality but a number of chemical contaminants (both inorganic and organic) found in water also causes great concern. Both bacteriological and chemical contamination of drinking water may have a short- or long-term deleterious effect on human health and nutrition.

Hence, drinking water should essentially be:

- Free from pathogenic (disease-causing) organisms
- Clear and transparent
- Not saline (salty)
- Free from offensive taste and smell
- Free from chemical contaminations that have adverse effect on human health

I. Chakravarty (✉)
Public Health Engineering Department, Water And Sanitation Support Organisation, Kiran Shankar Roy Road, Kolkata 700001, West Bengal, India
e-mail: indiracal@hotmail.com

K.J. Nath, V.P. Sharma (eds.), *Water and Sanitation in the New Millennium*,
DOI 10.1007/978-81-322-3745-7_19

- Free from chemicals that can cause corrosion to pipe and appurtenances or stains clothes/utensils, etc.

The BIS in India has developed such a standard. For microbiological assessment, an indicator organism is used which signifies the presence of pathogenic organism of concern and presence of faecal material. The organism needs to respond to natural environmental conditions and to treatment processes, and it can be isolated, identified and enumerated. The presence of a high ratio of indicator/pathogen is a point of concern, and the source it came from needs to be identified.

For chemical contaminants like arsenic, mercury, nitrates, fluoride, pesticides, etc., specific tests and identification methods are available which are standardized and are nationally as well as internationally accepted.

19.1.2 Water- and Sanitation-Related Infections [2]

There are various kinds of water- and sanitation-related diseases and as per the routes of infections they are classified as follows:

(i) Water-related infections:
  (a) Waterborne route
  (b) Water-washed route
  (c) Water-based route
  (d) Insect vector route
(ii) Excreta-related infections
  (a) Transmission via infected excreta
  (b) Transmission by an excreta-related insect vector
(iii) Water- and excreta-related infections

19.2 Interrelationship of WASH with Nutrition

This section highlights the basic facts on the interrelationship [3] between the WASH and its impact on the nutritional status of communities. Though most of the health indicators have been steadily improving over the years (NFHS-3) [4], but there has been hardly any improvement in the nutritional status of children.

Existing Scenario of Nutrition

There is unacceptable prevalence of undernutrition in the children, and the reasons for this possibly are:

- Inadequate intake of food
- Imbalanced diet
- Loss of nutrients from body due to infections
- Excessive physical labour

The last two causes are intricately related to unavailability of clean water, proper sanitation and proper personal as well as environmental hygiene.

The following data highlights the severity of the problem [4], during the 2005–2006 period as per the NFHS-3 survey:

- 42.5% of the children under the age of 5 years are underweight (low weight for age).
- 48% of the children are stunted (low height for age).
- 19.8% of the children are wasted (low weight for height).
- In poorer states, the situation is even worse with over 50% of children being underweight.

The later survey [19] in the year 2015–2016 (NFHS-4) also indicate not very significant improvement. Many of the children often start out at a disadvantage, as the national average of children born with low birth weight (<2.5 kg) is 22%. The situation does not improve very much for adolescents or adults. Thirty-six percent of adult women and 34% of adult men suffer from chronic energy deficiency (BMI < 18.5) with higher rates in rural and urban slum areas. In states like Bihar, Chhattisgarh and Jharkhand, the rates are over 40% [4].

Unsafe water and inadequate sanitation impacts on the nutritional status of the community in various ways as following:

19.2.1 WASH and Infection

Unsafe water, lack of sanitation and hygiene and unsafe food resulting from polluted water or poor sanitation and hygiene leads to **infections** like diarrhoea which cause loss of valuable nutrients from the body as well as disturbance in absorption from the gut; low assimilation and hampered metabolism. This eventually leads to undernutrition.

The malnutrition–infection cycle given below highlights these important linkages (Fig. 19.1).

Hence, unless a comprehensive approach is taken where WASH coverage is given along with adequate food, undernutrition cannot be controlled or overcome. Table 19.1 clearly indicates how maternal malnutrition impacts on the outcome of pregnancy, indicating a long-term impact [5].

Hence, a comprehensive approach to control undernutrition by provision of adequate and safe food as well as water will not only improve the health and nutritional status of the present generation but will also have a significant positive impact on the future generations [6].

19.2.2 Worm Infestation and Anaemia

Unsatisfactory sanitary facilities and open defecation result in **worm infestation** which may lead to **anaemia** as a precipitating factor as well as related undernutrition.

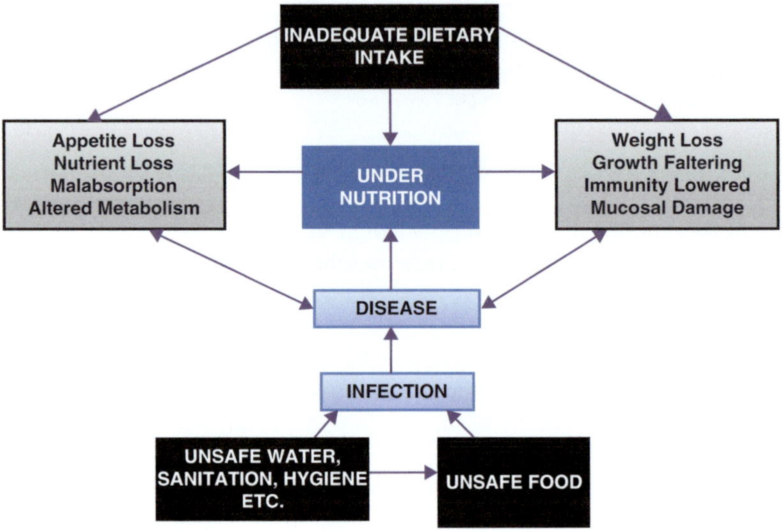

Fig. 19.1 Malnutrition–infection cycle [2]
Source: By Indira Chakravarty for PHED, GOWB report

Table 19.1 Maternal nutrition status and outcome of pregnancy

Group (according to nutritional status in childhood)	Foetal loss (%)	Birth weight (Kg.)	Birth weight below 2.5 Kg. (%)	Death during infancy (%)
Severely malnourished	11.8	2.41 ± 0.09	52.9	11.8
Moderately malnourished	8.9	2.57 ± 0.06	42.2	8.9
Mildly malnourished	8.2	2.55 ± 0.05	37.1	3.3
Normal	3.3	2.62 ± 0.08	38.3	6.9

Source: Women and Nutrition in India, Nutrition Foundation of India [7]

The rise in the prevalence rate of anaemia in the country has not been controlled satisfactorily in spite of the fact that specific national programmes on control of anaemia are in place. As a matter of fact, in some of the vulnerable groups, there actually is a rise in the prevalence rate of anaemia.

The NFHS-3 survey [4] highlighted widespread anaemia, with its prevalence actually increasing in some categories, such as in children between 6–59 months, where the rates increased from 74% in NFHS-2 (1998–1999) to 79% in NFHS-3 (2005–2006). Anaemia in women of reproductive age has also increased from 52% to 56% over this time period. 69% of boys and 70% of girls suffered from anaemia [4].

The more recent NFHS-4 survey [19] conducted during the two 2015–2016 years indicate no significant improvement in Anaemia status. The causes of anaemia are as under:

- Lack of intake of iron-rich food
- Intake of inhibitors affecting iron absorption in the body

Table 19.2 Effect of maternal haemoglobin status on birth weight and perinatal mortality

	Haemoglobin (g/dl)			
Parameters	<5.0	5.0–7.9	8.0–10.9	$\geq$ 11.0
No. of cases	312	443	1202	1783
Mean birth Wt. (g.)	2400 ± 810	2530 ± 651	2660 ± 544	2710 ± 496
Prematurity (%)	34.5	18.2	12.5	10.2
Birth weight (under weight babies <2500 g) %	62.3	38.4	26.4	23.1
Perinatal mortality (rate/1000 birth)	400.6	130.9	64.1	44.9

Source: Women and Nutrition of India, Nutrition Foundation of India [7]

- Worm infestation due to lack of sanitation
- Infections due to unsafe water or food, unhygienic environment, etc. causing loss of nutrients from the body as well as affecting absorption
- Any other causes which lead to loss of iron from body, e.g. excessive bleeding, child birth, etc.
- Specific physiological conditions like pregnancy

Many of these are due to unsatisfactory WASH situation.

It is a well-known fact that anaemia leads to low birth weight babies (LBW). The prevalence of LBW babies (weighing less than 2500 gms. at birth) is nearly unchanged since 1979. It is most pronounced in urban slums and rural areas. The Table 19.2 clearly shows the impact of anaemia on the incidence of LBW babies.

In many cases, lack of safe water and sanitation is a major cause of precipitating such situations. Hence, communities must have proper access to safe and hygienic water as well as food along with sanitation and hygiene to reduce anaemia levels in all sectors of population. Adequate and balanced food intake has to be covered by proper WASH inputs.

Table 19.3 compares a multistate data, as compiled in 2004 in the Human Development Report (West Bengal) supported by UNDP [8] by the author to indicate the usefulness of a comprehensive approach. Orissa having least coverage of latrines and access to safe water has maximum anaemia though other factors related to food intake or ANC are not much different as compared to those in other states.

19.2.3 Accessibility to Water and Nutrition

Unavailability of water near habitats results in long walks (particularly for the girl child and women who are the major carriers of water). This eventually leads to expenditure of valuable nutritional energy (calories), from the body. It also results in loss of time, which otherwise could have been used more fruitfully in caring for children, economic benefits (earning), or for resting (reduction of drudgery). Hence, inaccessibility to water leads to **loss of both calorific energy as well as time**. This eventually leads to malnutrition and ill health.

Table 19.3 A Statewise correlation of anaemia in adult females with other related factors

States/India	Anaemia (mod + severe anaemia)	BMI <18.5	Mean iron intake (mg/day) [RDA-30]	Vit. C intake	Tuber intake (gm/day)	Pulses intake	Latrine availability	Total literacy	Access to safe drinking water in households	% who received all recommended types of ANC
Orissa	63.0	46.0	118.8 ± 8.5	62 ± 68	78	16 ± 22	2.9	37.7	64.1	21.4
W.B.	62.7	45.9	11.6 ± 65	68 ± 48	132	9 ± 16	18.1	40.4	88.5	19.7
M.P.	54.3	41.9	15.5 ± 12.7	33 ± 30	42	29 ± 31	6.4	47.1	68.4	10.9
A.P.	49.8	42.0	9.7 ± 6.9	62 ± 68	27	22 ± 31	12.2	31.7	80.1	35.6
Gujarat	46.3	33.1	31.5 ± 17.9	42 ± 43	94	40 ± 34	60.1	57.7	84.1	25.0
Kerala	22.7	18.2	9.3 ± 3.8	23 ± 30	72	15 ± 27	90.6	85.4	23.4	64.9
Pooled/India	51.8	39.4	16.2 ± 12.8	36 ± 43	52	28 ± 31	26.2	46.9	78.0	20.0

Intakes have been shown for moderately active NPNL Women (NNMB Report, 2002)
Source: Compiled by Indira Chakravarty for HRD (2004) [8]

19.2.4 Chemical Pollution of Water and Nutrition

Pollutants in water like arsenic, fluoride, etc. lead to a direct **toxic impact on health and nutrition**. Moreover, arsenic in particular has been reported to enter the food chain at several points causing toxic implications through food as well. Pollution occurs due to use of contaminated water not only for drinking but also for agriculture, cooking, etc.

19.3 Case Studies on Impact of WASH on Nutrition

19.3.1 Impact of WASH-Related Interventions on Nutritional Status

A WHO supported study [9] done in rural areas of Medinipur district in West Bengal indicated that upgrading existing water sources and provision of new ones along with adequate sanitation facilities linked to health education significantly improved the nutritional status of the children. No food supplementation was provided. The WASH inputs provided were:

- Construction of pit latrines (community type)
- Construction of tube wells
- Construction of dug wells
- Renovation of dug wells

The targeted IEC strategies adopted for health and hygiene education were:

- House to house
- With help of front line workers like Panchayats, ICDS Centres and local health centre along with local NGOs
- Constant monitoring and linked awareness generation

The impact on diarrhoeal diseases and worm infestation was most significant as the rate of diarrheal diseases decreased from 27.27% to 21.06% and that of worm infestation from 85% to 75.35% (Ref. WHO Study, Indira Chakravarty).

As a consequence, after a year, there was a significant improvement in the malnutrition scenario as given in Table 19.4.

Table 19.4 Change in nutritional status of children (weight for age) after provision of WES [20]

Nutritional status	Baseline (%)	Post-intervention
Normal	5.27	13.46
Grade I	26.31	38.46
Grade II	57.89	38.46
Grade III	10.83	9.62

Source: By Indira Chakravarty, WHO supported study report

In conclusion, it can therefore be inferred that there can be significant improvement in height and weight for age (immediate nutritional status) as well as height for age (post-nutritional status) after provision of adequate and safe water and sanitation linked to health and hygiene education.

19.3.2 Impact of WASH on Food Safety

It is a well-known fact that unsafe food and water leads to infections like diarrhoea, dysentery, etc. which eventually result in causing undernutrition and ill health. Street foods are a major source of nutrition for the urban population and particularly for the urban poor [10, 11]. It is most suitable in the urban conditions because of the following facts:

- Nutritious and well balanced
- Cheapest source of nutrition
- Wide variety
- Easy accessibility
- Requires minimum space
- Very tasty and fresh
- Source of earning for unemployed
- Place of social interaction
- Caters to all strata of population

However, all these positive points get compromised as it is also a source of major infections (high contamination with faecal coliforms and other pathogens) due to the following reasons [12–14]:

- Poor handling
- Poor quality of water
- Poor sanitation
- Poor garbage disposal facilities
- Poor storage conditions

Hence, food safety in this case which affects lives of millions of people is jeopardized due to unclean water, poor handling, poor personal and food hygiene and poor environmental conditions. Hence, the major inputs needed for such food establishments are availability of clean water and sanitation proper hygiene practices by vendors and cleanliness all around [15–18].

19.3.3 Impact of Easy Access to Water on the Nutritional Status

Increased accessibility to clean water saves time and energy (calories) loss from the body. A comprehensive investigation was conducted with UNICEF support [9] by the author covering 24 villages in different regions of Nepal, **viz.** mountainous, hilly and terai regions. The parameters investigated were:

- Epidemiological information
- Nutritional status
- Dietary intake
- Health parameters
- Water collection, usage, quality, etc.
- Time spent for water collection
- Energy spent for water collection
- Adverse effects of carrying water in hilly and mountainous zones
- Housing/sanitation, etc.
- Socio-economic status
- Utilization of saved time – social, economic, child care, etc.

It was noted that if water sources are provided near habitats, then it saved the women a walk for about 3–4 h or more to carry back water from the source to home. This also resulted in saving of about 20% of body energy (calories). Hence, provision of adequate water at an accessible distance from home will help in preventing undernutrition (Table 19.5) in the following ways:

- Preventing loss of valuable energy (calories) from body.
- Giving more free time to women to take care of children and family, increase income by taking up income generation activities, and getting physical rest (reduction in drudgery). Therefore, in such cases, it will be far more cost-effective and beneficial to create water sources near habitats instead of giving regular nutritional supplementation, as shown in following Tables 19.5 and 19.6 [9].

Table 19.5 Average time and energy saved

Description	Saved		
	Cross-sectional*	Longitudinal**	Mean
Energy (K. Cal)	588 (20.5%)	510 (18.08%)	549 (19.29%)
Time (Hrs.)	4.50	3.00	3.75

*Water already provided (compared with other villages without provision of water supply)
**Compared before and after provision of water to control villages
Source: By Indira Chakravarty, WHO supported study report

Table 19.6 Improvement in Nutritional Status of Children[a] (Weight for Age)

Status	Nutritional status of children (%)			
Pooled sample	Normal	Grade I	Grade II	Grade III
Control	20.4	32.7	26.6	20.4
Experimental	26.3	34.2	26.3	13.2

Significant improvement of nutritional status
[a]0–5 years
Pooled = Central, Midwest, Far West
Source: By Indira Chakravarty, WHO supported study report

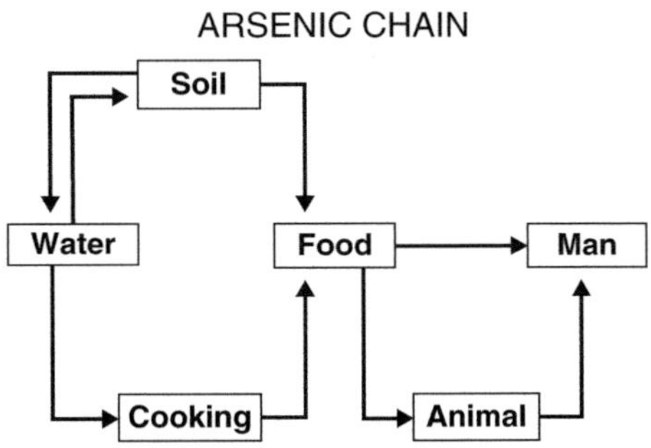

Fig. 19.2 Proposed canals to interlink Indian rivers

19.3.4 Impact of Arsenic Polluted Water on Nutritional Wellbeing

Entry of pollutants into food chain through contaminated water has a significant impact on human health and nutrition. A suitable example of entry of pollutants into food chain through contaminated water is the arsenic chain. The arsenic chain is active in the following manner (Fig. 19.2).

A study conducted with support of the Ministry of Health and Family Welfare, Government of India, in Malda district of West Bengal, indicated following facts [15]

- There is a significant level of entry of arsenic into the body through agricultural products being cultivated using arsenic contaminated water, e.g. grains, vegetables, fruits, etc.
- Food cooked using arsenic-contaminated water also absorbs a very high level of arsenic.
- The total entry of arsenic into the body through the food chain may account for more than 50% of arsenic entry load per day.
- In arsenic contaminated areas the levels of arsenic is higher in the entire chain as compared to that in the control areas, viz. water, soil, foods, nail, and urine.

19.4 Conclusion

Studies conducted over the past several years have consistently indicated that the rate of undernutrition in the country is going up in most cases, in spite of several national programmes being in place.

One of the major causes for this is perhaps the lack of inter-sectoral coordination where the importance of safe water, adequate sanitation as well as hygiene was not considered as the most necessary supportive factors during planning and implementation of the projects.

Based on study conducted by the author over several years clearly signify that Swachhta (cleanliness) in all respects is an essential necessity to attain health for all. [21, 22]

Acknowledgements The author is grateful to Prof. V. P. Sharma; NASI for his kind invitation to present this paper. The author would like to extend her thanks to WHO, UNICEF and FAO of the United Nations for supporting the case studies reported.

The author also thanks Public Health Engineering Department, Government of West Bengal, and the Foundation for Community Support and Development for their support.

References

1. Blossner M, de Onis M (2005) Malnitrition – quantifying the health impact at national and local levels, Environmental burden of disease series no. 12. World Health Organization, Geneva
2. Indira C et al (2010) Linkage of safe and adequate water, sanitation and hygiene with health, nutrition and related issues. A Publication by Public Health Engineering Department, Govt of West Bengal
3. Prüss-Üstün Bos AR, Gore F, Bartram J et al (2008) Safer water, better health: costs, benefits and sustainability of interventions to protect and promote health. Geneva: World Health Organization SAARC Disaster Management Center. 2008. South Asia Disaster Report SDMC, 2007. Ch. 11. Available at: http://saarc-sdmc.nic.in/pdf/publications /sdr/cover.pdf
4. National Family Health Survey 3, Ministry of Health and Family Welfare, Govt of India, International Institute for Population Science (IIPS) and Macro International (2007) International Family Health Survey (NFHS-3), 2005–2006: India, vol 1. IIPS, Mumbai
5. Von Braun J, Marie R, Ashok G (2008) The Facts: child Malnutrition in India. International Food Policy Research Institute
6. Gunther I, Fink G (2010) Water, sanitation, and children's health: evidence from 172 DHS surveys, World Bank Policy Research Working Paper No.5275
7. Gopalan C, Suminder K (1989) Woman and Nutrition of India, Nutrition Foundation of India
8. Indira C (2004) Human Development Report (UNDP) – West Bengal, Chapter on health and nutrition
9. Indira C (1995) Time energy saving after provision of safe water in Nepal, UNICEF Study
10. Winarno FG (1995) The regulatory and control aspects of street foods, FAO, Calcutta, India, (6–9 November)
11. Indira C (2011) Framework of action to ensure safety of street foods for implementation in the countries of South-East Asia, WHOSEARO
12. Indira C (1995) Working paper, Street foods in India, food inspection (Microbiological & Sanitary), Street Food Project in Calcutta. FAO, Calcutta
13. Indira C, Canet C, FAO (1996) 17/18, Food, nutrition and agriculture, p 30
14. Indira C (2007) A pilot study to assess the safety of street foods in India. Ministry of Food Processing Industries, Govt. of India

15. Indira C (2002) Arsenic toxicity and its impact on food chain. WHO South East Asia, Advisory Committee on Health Research
16. Indira C (2009) Final report on training and motivation of Vendors/consumers to improve the management of street foods in Chennai, Burdwan and Howrah, WHO, Sticker No: SE/08/032891
17. WHO Study; Assessment and improvement of street foods situation in municipal town of West Bengal, Burdwan, Foundation for Community Support and Development, 2001–2002
18. WHO Study (2009) Towards creation of Howrah as a Safe Food Town – A multi stakeholder approach; Foundation for community support and development
19. National Family Health Survey 4, Ministry of Health and Family Welfare, Govt of India, International Institute for Population Science (IIPS) and Macro International, International Family Health Survey (NFHS-4), 20015–16: India, vol 1. IIPS, Mumbai, Survey Conducted in 2015–2016
20. Indira C (1995) Time energy saving after provision of safe water in Nepal; UNICEF Report; Nepal
21. Indira C, Tanvir A (2016) A synthesis report of Bangladesh and India – exploring impacts of drinking water and sanitation on health and nutrition and socio- economic status of communities; ecosystem for life: a Bangladesh-India Initiative. International Union for Conservation of Nature (IUCN), Dhaka, Bangladesh, pp 1–68. www.iucn.org/E4L
22. Indira C (2017) Comparative assessment using multi – sectoral public health indicators in various areas of Tamil Nadu. State Planning Commission, Tamil Nadu; Foundation for Community Support and Development; 2250/SPC/HSW/2015-16; 2016, pp 1–258 (In Press)

Prof. Dr. Indira Chakravarty has the highest qualification in science – **PhD and DSc.**

She is a **Padma Shri Awardee.**

Some of the other **prestigious awards** received by her are **the Eduardo Sauma Global Award from FAO of the United Nations; President's Global Leadership Award in Public Health from the University of South Florida, USA; Indira Gandhi Priyadarshini Award; etc.**

She is a **fellow** of many renowned organizations and has **over 250 publications to her credit.**

Presently

Dr. Chakravarty is chief adviser of the Public Health Department, Government of West Bengal; **Director** (independent) of ECL, Coal India; **Chairman** of the Water and Beverage Scientific Panel, Food Safety and Standards Authority of India (FSSAI), Ministry of Health and Family Welfare, Government of India; **Chairman** of RMRC, ICMR, Department of Medical Research, Ministry of Health and Family Welfare, Government of India; **Chairman** of the Stop Diarrhoea programme, Save the Children; **Chairman** of the Innovation Center, Schevaren; **Member** of Hygiene Index, Reckitt Benckiser; **Member** of the Core Committee, WASH, Swachh Bharat, Ministry of Drinking Water and Sanitation, Government of India; and **Member** of the International Board of Advisers, International Institute for Global Health, United Nations University (UNU).

Formerly

She was **Additional Director General of Health Services**, **Director** of All India Institute of Hygiene and Public Health, Ministry of Health and Family Welfare, Government of India; **Director** of Chittaranjan National Cancer Institute, Ministry of Health and Family Welfare, Government of India; **Dean and Head of the Department** of Biochemistry and Nutrition, All India Institute of Hygiene and Public Health, Government of India; and **regional adviser** ac (Nutr) of WHO, Southeast Asia Regional office and Regional Director, MI, IDRC (South Asia).

Save Ganga Movement: A Gandhian Non-Violent Movement For A Non-Violent Culture Of Development

20

Rama Rauta

A Gandhian Non-violent Movement for a Non-violent Culture of Development

Gandhi symbolizes a culture of Truth and non-violence, i.e. *a culture of pursuit of ethical perfection as the ultimate goal of life and pursuit of selfless ethical life of universal love as means; the Ganga symbolizes all rivers and water bodies and the Giriraj Himalaya symbolizes all mountains, forests and wildlife.*

Report on Save Ganga and Save Himalaya Meeting-cum-Panel Discussion on 12 March 2015 at Gandhi Darshan, New Delhi.

Topic of discussion: What must we do to save the Ganga and the Himalayas in the context of the present extremely eco-hostile, out-and-out consumerist global market culture of unlimited desires?

Participants

- Sushri Uma Bharti, Hon'ble Minister of Water Resources, River Development and Ganga Rejuvenation, was the Chief Guest of the function
- Smt. Rama Rauta, Founder and Convener, Save Ganga Movement chaired the function

R. Rauta (✉)
Save Ganga Movement, Pune, India

National Women's Organization, Pune, India
e-mail: ramarauta@rediffmail.com; ramarauta29@gmail.com

K.J. Nath, V.P. Sharma (eds.), *Water and Sanitation in the New Millennium*,
DOI 10.1007/978-81-322-3745-7_20

- Prof. Vinod Tare, Indian Institute of Technology, Kanpur and Co-ordinator, Ganga River Basin Management Plan gave the keynote address.
- Parampujya Swami Nikhilanandaji, Regional Head, Chinmaya Mission, New Delhi
- Parampujya Acharya(Dr.)Sri Lokesh Muni, President, Ahimsa Vishwa Bharti, New Delhi
- Revered Imam Umer Ahmed Ilyasi, Chief Imam, All India Imam Organization, New Delhi
- Parampujya Swami Bhakt Rasmrita, ISCKON, Mumbai
- Rev. Fr. Dominic Emmanuel
- Shri Paramjeet Singh Chandok
- Revered Acharaya Vivek Muni
- Dr. Mohan Singh Rawat Gaonwasi, Former Minister of Uttrakhand, Expert Member NGRBA
- Shri M.C. Mehta, Noted Environmentalist and Supreme Court Advocate
- Shri Vinod Kumar Agrawal, Mokshda, New Delhi

Many other distinguished persons from different streams of life were also present in this Save Ganga and Save Himalayas Meeting.

The following Save Ganga and Save Himalayas Recommendations were discussed and accepted in the Save Ganga and Save Himalaya Meeting-cum-Panel Discussion and presented to Sushree Uma Bharatiji, Hon'ble Minister of Water Resources, River Development and Ganga Rejuvenation, who was the Chief Guest of the function.

The Charter of Save Ganga and Save Himalayas Recommendations

1. The Ganga must be constitutionally declared as the National River with statutory provisions that ensure due respect and protection to her, considering her National River status analogous to the 'Prevention of Insults to National Honour Act' for the national flag and national anthem.
2. We must accept the recommendation of the consortium of 7 IITs that:

   (A) In place of the present policy of allowing treated sewage into our rivers, we must adopt the policy of zero discharge into the river and promote reuse and recycle of waste water after proper treatment (tertiary-level treatment).

   (B) Industrial effluents and hospital wastes, treated or untreated, must never be allowed to enter into the rivers and must not also be allowed to mix with the sewage, which should be converted into valuable manure for organic farming, industries must treat their effluent and use recycled water.

   (C) Organic farming should be promoted in a massive way for decreasing the non-point sources of pollution of rivers such as hazardous chemicals from agricultural run-off into the rivers and also for maintaining soil fertility, checking the groundwater degradation, reducing water requirement of crops, producing health-friendly food, etc.

[Since our rivers are the source of drinking water for crores of our common people and also for the animals and STPs cannot convert sewage into potable water, our

Save Ganga Movement has been demanding since long time that we must adopt throughout our country the policy of zero discharge into the rivers which is also now recommended by the consortium of 7 IITs which is preparing National Ganga River Basin Management Plan (NGRBMP).]

3. The bactericidal, health-promoting, non-putrefying and self-purifying properties of the water of Ganga should be restored and conserved. Scientists claim that Ganga has the unique quality of self-purification capacity due to the presence of high level of bactericidal copper and chromium and perhaps of uranium and thorium at the Himalayan sediments and different types of beneficent bacteria and phages in the sediments of the river which kill the harmful bacteria coli forms.
4. Treatment of the sewage through 'pond system and plant-based management of sewage and waste treatment' and using nutrient-rich treated waste water for organic farming is the cheapest and durable method and needs least management and electricity.

A massive time-bound plantation programme on the banks of the river Ganga from Gangotri to Ganga Sagar, along with the development of constructed wetlands for sewage treatment in major cities on the banks of the river wherever possible, should be undertaken with the help of NBRI, Lucknow, NEERI, Nagpur, along with other prominent research centres of environmental science/botany/engineering from our universities/colleges and various like-minded NGOs and local people. Herbal strip along the river should be promoted under the scheme of Rural Eco-friendly River Front Development of the Ganga.

The eight-decade-old East Kolkata Wetland constitutes an ideal example of a system of natural biotreatment of urban waste water through 'Pond System and Plant Based Management of Sewage Treatment'. Recycling and utilizing the treated waste water for fish culture and agriculture: it provides about 13,000 tons of fish per year obtained from about 300 waste water fed ponds, 150 tons of fresh vegetables per day from the small-scale horticulture plots irrigated with the treated waste water, water for irrigating paddy cultivation and livelihood for about 50,000 common people and also serves as a natural sponge absorbing excess rainfall.

5. The highly earthquake-prone, eco-fragile and ecologically, aesthetically and religiously invaluable Uttarakhand region of the Ganga Basin must be declared 'ecologically fragile and a sanctuary for Himalayan flora and fauna' and its rivers 'wild rivers'. All steps must be taken to preserve its rivers and vegetation in pristine condition. It would be a major step towards the realization of our National Mission of saving the Himalayan Ecosystem as a part of the National Action Plan for Climate Change and also of the National Action Plan for Preservation of our Biodiversity.

Since crores of our people since ages consider the entire Himalayan region of the Ganga with all its tributaries to be the zone for self-purification and spiritual enlightenment (TapoBhumi and Adhyatma-Bhumi), we should also

declare this religiously invaluable Uttarakhand region to be our national Spiritual Heritage Zone.

We must have a high-powered National Himalayan Ecology Preservation and Restoration Authority headed by Hon'ble Prime Minister to save the highly fragile invaluable ecology of the young Himalayas. The impending catastrophe of fast receding of Himalayan glaciers has to be understood and tackled at a regional and global level. At the regional level, it must involve all Himalayan nations. India should take a major global initiative in this direction to tackle this regional and global crisis.

6. To begin with, we must make Uttarakhand an absolutely eco-friendly ideal Himalayan state and must take time-bound decisive steps to make the Yamuna at Delhi completely and permanently free from pollution, which would set an example for the entire country. **An adequate flow of natural fresh water must be allowed to flow on the Ganga bed and the Yamuna bed throughout the stretch of the rivers throughout the year not only to protect and preserve their ecology but also to meet the basic water needs of the cities, towns and villages situated on their banks and restore their self-purifying capacity**. At present in dry season, the three large barrages at Haridwar, Bijnor and Narora divert 100% of the river's water into its canals, and the Ganga is totally bereft of Gangajal after the Narora barrage. It is highly deplorable that our national capital Delhi is the greatest polluter of the river Yamuna, the largest tributary of our National River Ganga. In dry season, no water is allowed to flow in the Yamuna River downstream to Hathni Kund barrage in Haryana, and what reaches the holy cities of Mathura and Vrindavan is mainly the treated or untreated domestic and industrial waste water contributed by various drains joining the Yamuna at Delhi.

Since the Ganga is our national river and crores of people consider her to be Devine Mother, at least the main stream of the Ganga must be maintained close to its pristine and natural state.

7. The consortium of IITs, which is preparing the National Ganga River Basin Management Plan(NGRBMP), should make a holistic comprehensive scientific study of (a) the problem of construction of extremely eco-hostile dams for hydropower in the highly earthquake-prone, eco-fragile Himalaya region of the Ganga Basin, (b) the problem of construction of extremely eco-hostile barrages for Navigation in the River Ganga and (c) the problem of extremely eco-hostile interlinking of the rivers within the Ganga basin and between the rivers of the Ganga basin and the Peninsular rivers: **it has not made any study of these issues so far. Since such scientific study of the issues is necessary for preparing a holistic comprehensive GRBMP, the government should provide all the necessary help to the consortium of 7 IITs to make such studies of the issues freely with dignity as soon as possible. Decision on these controversial issues must be postponed till a national consensus on**

these issues is available on the basis of holistic scientific knowledge about the short-term as well as long-term harmful as well as beneficial consequences of such undertakings. We must look beyond short-term economic benefits and have a holistic scientific study about the long-term environmental cost of the projects whose victims would be mainly, in addition to the adversely effected local people, our future generations and our dumb and deaf fellow creatures. What we need to ensure our lasting water security is decentralized basin restoration, recharging and management approaches that consider a host of small and medium ecologically sustainable measures involving participation of local people.

8. No encroachment should be allowed on either side of the banks of Ganga at least within 200–300 metres. Construction of permanent structures for residential, commercial or industrial purposes in the active flood plains of any river must be prohibited.
9. We must have a law protecting River rights and River Guards to prevent crimes against rivers and River Courts to try crimes against rivers.
10. There should be disincentives in the form of proper fines to the states in the Ganga basin in proportion to the quantity and quality of pollution a state adds to the river in the state. **The NRCD must have a monitoring mechanism to regularly monitor the water quality of the rivers at the entry and exit points of each state**.

Similarly, there should be disincentives in the form of proper fines to the cities/towns in proportion to the quantity and quality of pollution a city/town has added to the river which flows through or near it. There should be a monitoring mechanism to monitor regularly the water quality of the rivers at the entry and exit points of each major polluting city/town.

11. The National Ganga River Basin Authority (NGRBA) must take the full responsibility of the protection of our national river. Since the Ganga flows through many states, it would be the best course if NGRBA takes the full responsibility of making and keeping the Ganga and its tributaries completely and permanently free from pollution through time-bound steps, leaving no scope for the central and state government authorities blaming each other for the failures.

The expert members of NGRBA must be given due role and importance in this organization both in decision-making and in implementation. NGRBA must have transparency and accountability in every sphere of its activities.

12. **There is no scarcity of money, knowledge and skills with us to save our rivers including the Ganga. There is lack of will due to our moral bankruptcy. Environmental ethics should be taught as a part of the syllabus on ethics which must be taught as a compulsory subject, both at the school as well as at the college level.** Teaching environmental ethics without discussing various fundamental questions concerning ethical values, the value and means of an ethical life would be of little significance. **We must study critically the views of great religions and of great teachers and thinkers of mankind about vari-**

ous fundamental issues of ethics concerning ethical values and the value and means of ethically good life, which would be a major step in the direction to overcome our present deep rooted moral and spiritual crisis.

It is our deep conviction that acceptance and implementation of the above 'Charter of Save Ganga and Save Himalayas Recommendations' is necessary to rejuvenate and preserve our national River the Ganga and her tributaries. They are essentially based on: (**a**) 'The Charter of 7 Save Ganga and Save Himalayas Recommendations' made at the Gandhi Jayanti Save Ganga and Save Himalayas Function held at Bhartiya Vidya Bhavan, Mumbai, on 2 October 2013; (**b**) 'The Save Ganga and Save Himalayas Resolution', passed at the Save Ganga and Save Himalayas Function held at Gandhi Darshan, Rajghat, New Delhi, on 12 March 2011 and 12 March 2012; (**c**) 'Save the Rivers of Maharashtra Resolution' passed in the seminars held on the occasion of Punyatithi of Mahatma Gandhi at Gandhi National Memorial, Agakhan Palace, Pune, on 30 January 2011, 2012, 2013 and 2014; and (**d**)'The Charter of Ten Demands to save the Ganga and the Himalayas' accepted by the former Hon'ble Prime Minister Dr. Manmohan Singh on 12 March 2009. (These are available in our website, www. Savegangamovement.org.)

***

In the long term, we must make serious efforts to solve the problems of eco-hostile industrialization, urbanization and population-growth, which constitute the root of our problem of environmental degradation in general and the slow death of our rivers in particular, by a radical change from our current ideas of development and growth to a Gandhian alternative, which essentially involves retelling the basic ethics of all great religions in the context of our technological age and which is the surest and perhaps the only solution to our impending catastrophic global ecological crisis, including the problem of global warming and climate change. Industrialization and urbanization must be in the service of rural India, and development must be non-violent with loving care of our fellow lower form of creatures and life- and health-sustaining natural systems. We must not allow our population to grow beyond the limit which would cause great harm to our future generations or to our fellow lower form of creatures, and if it has gone beyond the limit, we must try to bring it down in some ethical way to the desirable level. We must take time-bound decisive steps to solve the problem of forced migration of rural unemployed to cities by converting rural India into ideal places for leading enlightened ethical life of simple living and high thinking with dignified exhilarating and health-giving body labour in an atmosphere of natural purity and beauty. **We must root out corruption, which is proving to be the biggest hurdle in the way of implementing our various governmental programmes of public services, with the help of good governance and character building ethics education**. Ultimately, we must create a new paradigm of development for India based on the Gandhian principles of Truth and non-violence (for a discussion on this issue, please see the note entitled 'Gandhian Solution to Global Ecological Crisis' in our website, www. Savegangamovement.org.).

The great teachers and seers of our ancient Indian civilization explicitly accept universal non-violence, i.e. non-violence to both human and non-human life, to be the foundation of ethics. They see clearly that a life of perfect enlightened selfless ethical universal love and renunciation constitutes the core of Truth, i.e. of the true ultimate goal of life, and pursuit of selfless ethical life of love, serving selflessly to society to the best of one's ability through some work required for the general good and making constant effort to progress towards ethical perfection, constitutes the core of its means. They see clearly that enlightened selfless desireless ethical life of love (i.e. enlightened selfless ethical life of love with detachment to the fruits of actions) is intrinsically peaceful and blissful and that a liberated life is eternally the best form of life: the happiness which intrinsically involves in such a life is of the highest kind and everlasting. They see clearly that any person through conscious effort can pursue liberation and progress towards it from evil to good life, from selfish good life to unselfish good life, from unselfish good life to enlightened selfless good life and from it finally to liberated life. They see clearly that such a human life involves selfless ethical service to humanity and its fellow creatures to the best of one's ability in the best possible way, whether the contribution it makes for the betterment of the lifeworld is very high or very low. They explicitly accept that pursuit of wealth and pleasure within the limits of ethics is essential not only for the pursuit of the ultimate goal of life but also for lasting development, prosperity, peace and happiness in society. Gandhiji and Swami Vivekananda see clearly that all great religions also explicitly or implicitly accept the same.

All great religions, explicitly or implicitly, accept the natural ecosystems which sustain life and provide various essentials of life freely to masses to be sacred and invaluable. Neither the great seers of our Indian civilization nor the seers of various great religions would approve the present fundamentally unspiritual, extremely eco-hostile, inherently unsustainable and out-and-out consumerist global market culture of unlimited desires for wealth and luxuries which has caused disappearance of tens of thousands of plant and animal species and continues to cause greater and greater violence to our life- and health-sustaining natural systems through unethical use of science and technology, which has devastating implications in the long run for our future generations as well as for the entire life world. According to the 2014 Living Planet Report of WWF, our planet earth lost 52% of its wildlife in past 40 years between 1970 and 2010 (The Times of India, Mumbai, October 1, 2014). They would agree with the Gandhian view that the modern civilization's culture of having unlimited desires and going to the ends of the earth with the help of science and technology causing great irreparable harm to earth's life- and health-sustaining natural systems in search of their satisfaction is satanic and suicidal. Saving the Ganga and its tributaries and their source the Himalayas and the Godavari and Krishna and their source, the Western Ghats is absolutely necessary for us to have a lasting sustainable development: it is necessary to ensure clean water, food, air and health security to our masses. Protection of the Ganga, symbolizing all rivers and water bodies, and the Giriraj Himalaya, symbolizing all mountains, forests and wildlife, must be accorded highest priority in our national development process. We must protect and take loving care of the delicate and holistic balance that exists in the ecosystems of nature which are invaluable from the view point of utility, aesthetics as well as religion, and try to restore wherever possible our degraded ecosystems. Surely the message of enlightened selfless desireless ethical action of universal love of the Indian civilization, which the Ganga has given to the world, has the potentiality to save the world from the impending catastrophic global ecological crisis along with its present deep rooted all pervasive moral and spiritual crisis. Let India, the land of many great religions, provide the world a culture of non-violence and Truth, i.e. the true ultimate meaning of life, in the context of our present technological age, where all religions could grow harmoniously in spite of their differences in the realm of metaphysics on the basis of the knowledge about their fundamental ethical unity and truth and the central place of ethics in religious life, where science and technology would be used for development only within the limits of ethics and where development would take place with loving care of the invaluable countless kinds of flora and fauna of our life giving and life-sustaining aesthetically and religiously invaluable natural systems.

Mrs. Rama Rauta is a Gandhian social activist devoted to the cause of saving the Ganga symbolizing all our rivers and water bodies and the Himalayas symbolizing all our mountains, forests and wildlife through a Gandhian non-violent movement. She is the founder and president of the National Women's Organization, Pune, and the founder and convener of the Save Ganga Movement. She is a former expert member of the National Ganga River Basin Authority (NGRBA), Government of India, headed by the then prime minister of India Dr. Manmohan Singh. She is at present a member of the Expert Advisory Committee, Ministry of Water Resources, River Development and Ganga Rejuvenation, Government of India.

MIX
Papier aus verantwortungsvollen Quellen
Paper from responsible sources
FSC® C105338

If you have any concerns about our products,
you can contact us on
ProductSafety@springernature.com

In case Publisher is established outside the EU,
the EU authorized representative is:
Springer Nature Customer Service Center GmbH
Europaplatz 3, 69115 Heidelberg, Germany

Printed by Libri Plureos GmbH
in Hamburg, Germany